NORM ESTIMATIONS FOR OPERATOR-VALUED FUNCTIONS AND APPLICATIONS

PURE AND APPLIED MATHEMATICS

A Program of Monographs, Textbooks, and Lecture Notes

MONOGRAPHS AND TEXTBOOKS IN PURE AND APPLIED MATHEMATICS

53. *C. Sadosky,* Interpolation of Operators and Singular Integrals (1979)
54. *J. Cronin,* Differential Equations (1980)
55. *C. W. Groetsch,* Elements of Applicable Functional Analysis (1980)
56. *I. Vaisman,* Foundations of Three-Dimensional Euclidean Geometry (1980)
57. *H. I. Freedan,* Deterministic Mathematical Models in Population Ecology (1980)
58. *S. B. Chae,* Lebesgue Integration (1980)
59. *C. S. Rees et al.,* Theory and Applications of Fourier Analysis (1981)
60. *L. Nachbin,* Introduction to Functional Analysis (R. M. Aron, trans.) (1981)
61. *G. Orzech and M. Orzech,* Plane Algebraic Curves (1981)
62. *R. Johnsonbaugh and W. E. Pfaffenberger,* Foundations of Mathematical Analysis (1981)
63. *W. L. Voxman and R. H. Goetschel,* Advanced Calculus (1981)
64. *L. J. Corwin and R. H. Szczarba,* Multivariable Calculus (1982)
65. *V. I. Istrățescu,* Introduction to Linear Operator Theory (1981)
66. *R. D. Järvinen,* Finite and Infinite Dimensional Linear Spaces (1981)
67. *J. K. Beem and P. E. Ehrlich,* Global Lorentzian Geometry (1981)
68. *D. L. Armacost,* The Structure of Locally Compact Abelian Groups (1981)
69. *J. W. Brewer and M. K. Smith, eds.,* Emily Noether: A Tribute (1981)
70. *K. H. Kim,* Boolean Matrix Theory and Applications (1982)
71. *T. W. Wieting,* The Mathematical Theory of Chromatic Plane Ornaments (1982)
72. *D. B. Gauld,* Differential Topology (1982)
73. *R. L. Faber,* Foundations of Euclidean and Non-Euclidean Geometry (1983)
74. *M. Carmeli,* Statistical Theory and Random Matrices (1983)
75. *J. H. Carruth et al.,* The Theory of Topological Semigroups (1983)
76. *R. L. Faber,* Differential Geometry and Relativity Theory (1983)
77. *S. Barnett,* Polynomials and Linear Control Systems (1983)
78. *G. Karpilovsky,* Commutative Group Algebras (1983)
79. *F. Van Oystaeyen and A. Verschoren,* Relative Invariants of Rings (1983)
80. *I. Vaisman,* A First Course in Differential Geometry (1984)
81. *G. W. Swan,* Applications of Optimal Control Theory in Biomedicine (1984)
82. *T. Petrie and J. D. Randall,* Transformation Groups on Manifolds (1984)
83. *K. Goebel and S. Reich,* Uniform Convexity, Hyperbolic Geometry, and Nonexpansive Mappings (1984)
84. *T. Albu and C. Năstăsescu,* Relative Finiteness in Module Theory (1984)
85. *K. Hrbacek and T. Jech,* Introduction to Set Theory: Second Edition (1984)
86. *F. Van Oystaeyen and A. Verschoren,* Relative Invariants of Rings (1984)
87. *B. R. McDonald,* Linear Algebra Over Commutative Rings (1984)
88. *M. Namba,* Geometry of Projective Algebraic Curves (1984)
89. *G. F. Webb,* Theory of Nonlinear Age-Dependent Population Dynamics (1985)
90. *M. R. Bremner et al.,* Tables of Dominant Weight Multiplicities for Representations of Simple Lie Algebras (1985)
91. *A. E. Fekete,* Real Linear Algebra (1985)
92. *S. B. Chae,* Holomorphy and Calculus in Normed Spaces (1985)
93. *A. J. Jerri,* Introduction to Integral Equations with Applications (1985)
94. *G. Karpilovsky,* Projective Representations of Finite Groups (1985)
95. *L. Narici and E. Beckenstein,* Topological Vector Spaces (1985)
96. *J. Weeks,* The Shape of Space (1985)
97. *P. R. Gribik and K. O. Kortanek,* Extremal Methods of Operations Research (1985)
98. *J.-A. Chao and W. A. Woyczynski, eds.,* Probability Theory and Harmonic Analysis (1986)
99. *G. D. Crown et al.,* Abstract Algebra (1986)
100. *J. H. Carruth et al.,* The Theory of Topological Semigroups, Volume 2 (1986)
101. *R. S. Doran and V. A. Belfi,* Characterizations of C*-Algebras (1986)
102. *M. W. Jeter,* Mathematical Programming (1986)
103. *M. Altman,* A Unified Theory of Nonlinear Operator and Evolution Equations with Applications (1986)
104. *A. Verschoren,* Relative Invariants of Sheaves (1987)
105. *R. A. Usmani,* Applied Linear Algebra (1987)
106. *P. Blass and J. Lang,* Zariski Surfaces and Differential Equations in Characteristic $p > 0$ (1987)
107. *J. A. Reneke et al.,* Structured Hereditary Systems (1987)

Additional Volumes in Preparation

NORM ESTIMATIONS FOR OPERATOR-VALUED FUNCTIONS AND APPLICATIONS

Michael I. Gil'
Institute for Industrial Mathematics
Beersheva, Israel

CRC Press
Taylor & Francis Group
Boca Raton London New York

CRC Press is an imprint of the
Taylor & Francis Group, an **informa** business

CRC Press
Taylor & Francis Group
6000 Broken Sound Parkway NW, Suite 300
Boca Raton, FL 33487-2742

First issued in paperback 2019

© 1995 by Taylor & Francis Group, LLC
CRC Press is an imprint of Taylor & Francis Group, an Informa business

No claim to original U.S. Government works

ISBN-13: 978-0-8247-9609-9 (hbk)
ISBN-13: 978-0-367-40162-7 (pbk)

Visit the Taylor & Francis Web site at
http://www.taylorandfrancis.com

and the CRC Press Web site at
http://www.crcpress.com

Library of Congress Cataloging-in-Publication Data

Gil', M. I. (Mikhail Iosifovich)
    Norm estimations for operator-valued functions and applications / M. I. Gil'.
        p. cm. — (Monographs and textbooks in pure and applied mathematics; 192)
    Includes bibliographical references (p. –) and index.
    ISBN 0-8247-9609-8 (hardcover : alk. paper)
    1. Operator-valued functions. 2. Estimation theory. I. Title. II. Series.
QA323.G48   1995
515'.73—dc20
                                                            95-32196
                                                            CIP

Preface

Operator-valued functions are a classical topic in mathematics and its applications. For example, semigroups (exponential functions) of operators play a valuable role in the theory of differential equations, the resolvent is central to spectral theory, etc.

One of the first estimates for the norm of a regular matrix-valued function was established by I. M. Gel'fand and G. E. Shilov in connection with their investigations of partial differential equations. But this estimate is not sharp. It is not attained for any matrix. The problem of obtaining a precise estimate for the norm of a matrix-valued function was discussed in the literature repeatedly. In the late 1970s I obtained a precise estimate for regular matrix-valued functions. It is attained for normal matrices. Later this estimate was extended to various classes of nonselfadjoint operators such as Hilbert-Schmidt's operators, operators which are represented as the sum of unitary and compact ones (quasiunitary operators), quasi-Hermitian ones (i.e. linear, generally unbounded operators with a completely continuous imaginary component), etc. Singular integral and integral-differential operators are examples of quasi-Hermitian operators.

On the other hand, Carleman in the 1930s obtained an estimate for the norm of the resolvent of finite dimensional operators and of operators belonging to the Neumann-Schatten's ideal. In the early 1980s sharp estimates for norms of the resolvent of nonselfadjoint operators of various types were established. These results improve and extend Carleman's estimate.

The above-mentioned estimates are applied to spectrum perturbations, and to the stability analysis of systems of ordinary and partial differential equations, delay equations, integrodifferential, and Volterra integral equations.

By means of the estimates for resolvents, the well known spectrum perturbation result for selfadjoint operators is extended to various classes of nonselfadjoint operators. In addition, Schur's and Brown's classical inequalities for eigenvalues of matrices are improved. We also give a spectral representation for the resolvent of a nonselfadjoint operator by the multiplicative integral with respect to a spectral measure. This representation generalizes the one for selfadjoint operators. It is interesting that for nonselfadjoint operators the multiplicative integral with respect to a spectral measure plays the same role as the additive one for selfadjoint operators.

One of the basic methods for investigation of solution stability is the Lyapunov functions (functionals) method by which many strong results are obtained. But finding Lyapunov functions is usually difficult. By the estimates for norms of operator-valued functions new stability criteria are obtained. They make it possible to avoid the construction of Lyapunov's functions in appropriate situations. By these criteria, stability conditions of various equations are established. In particular, in this way we obtained stability conditions for control systems with distributed and concentrated parameters, for autogenerators, for predator-prey ecological systems, etc.

This book is devoted to the above-listed results. The aim of the book is to provide new tools for specialists in functional analysis and stability theory. This is the first book that

i) presents a systematic exposition of estimations for norm of operator-valued functions and their various applications,

ii) demonstrates a new approach to spectrum perturbations. It allows us to obtain precise bounds for spectra of nonselfadjoint operators,

iii) deals with new stability criteria for various classes of equations. They allow us to avoid the constructions of Lyapunov's functions in appropriate situations.

This book is directed not only to specialists in functional analysis and stability theory, but to anyone interested in various applications who has had at least a first-year graduate-level course in analysis. The functional analysis is developed as needed.

I was very fortunate to have had fruitful discussions with Professors M. A. Aizerman, I. S. Iohvidov, V. B. Kolmanovskii, M. A. Krasnoselskii, A. V. Kugel, A. D. Myshkis, A. Pokrovskii, B. N. Sadovskii and A. A. Voronov, to whom I am very grateful for their interest in my investigations.

It was very difficult to write the book in a language which I learned rather late in the day. It became possible due to the help of Professor Wolf Plotkin, and I wish to express my thanks to him.

I am particularly indebted to the editorial staff of Marcel Dekker, Inc. for their assistance.

It may safely be said that this book would never have been completed without the patience and encouragement of my wife, Lubov Gil'.

Michael I. Gil'

Contents

NORM ESTIMATIONS FOR OPERATOR-VALUED FUNCTIONS AND APPLICATIONS

Chapter 1

Matrix-Valued Functions

This chapter is devoted to matrix-valued functions. The main tool for norm estimates of matrix-valued functions in the nonnormal case is the quantity $g(A)$ for a matrix A, which captures the complications, coming from an intricate Jordan structure for A, completely in terms of the matrix entries of A.

In Section 1.1 we recall some basic notions of matrix theory.

In Section 1.2 the main results of the chapter (Theorem 1.2.1 and Theorem 1.2.2) are presented. Namely, estimations for the norm of a regular matrix-valued function and the resolvent of a matrix are considered. The proofs are given in Section 1.3 and in Section 1.4. In particular, in Section 1.3 an estimation for the powers of a nilpotent matrix is derived.

In Section 1.5 we establish inequalities between eigenvalues and singular numbers.

The multiplicative structure of the resolvent of a matrix is treated in Section 1.6. Namely, we obtain a representation of the resolvent by a product of simple matrices which extends the spectral representation of the resolvent of a normal matrix.

By the multiplicative representation for the resolvent, in Section 1.7 a

1

relation between the determinant and the norm of the resolvent is established.

1.1   Notation

Let $\mathbf{C}^n$ be a complex Euclidean space with the Euclidean norm

$$\|x\| = (|x_1|^2 + \ldots + |x_n|^2)^{1/2}$$

and the scalar product

$$(x, y) = \sum_{k=1}^{n} x_k \overline{y}_k$$

for $x \in (x_k)$, $y = (y_k) \in \mathbf{C}^n$.

For an $n \times n$-matrix (a linear operator in $\mathbf{C}^n$) A, its conjugate transpose is denoted by A^*,

$$\|A\| = \sup_{x \in C^n} \frac{\|Ax\|}{\|x\|}$$

is the norm of A,

$$\lambda_k(A);\ k = 1, \ldots, n$$

are eigenvalues of A including their multiplicities, $\sigma(A)$ is the spectrum of A, i.e.

$$\sigma(A) = \{\lambda_k(A)\}_{k=1}^{n}.$$

A complex number λ is *a regular point of A* if it does not belong to the spectrum of A, i.e. if $\lambda \neq \lambda_k(A), k = 1, \ldots, n$.

As usually, $det A$ denotes the determinant of A :

$$determinant(A) = det A = \prod_{k=1}^{n} \lambda_k(A).$$

It is well known that $det(AB) = det(A)det(B)$ for every matrix A and B.

The trace of A is denoted by $Tr A$:

$$Trace A = Tr A = \sum_{k=1}^{n} \lambda_k(A).$$

Recall that $Tr(AB) = Tr(BA)$ and

$$Tr(A + B) = Tr(A) + Tr(B)$$

for every matrices A and B.

Let I be the unit matrix. Everywhere below A^{-1} is the inverse matrix of A, and

$$R_\lambda(A) \equiv (A - \lambda I)^{-1},$$

where λ is a regular point of A. That is, $R_\lambda(A)$ is the resolvent of A.

Let $f(\lambda)$ be a scalar-valued function which is analytical in a neighborhood of $\sigma(A)$. We define the function $f(A)$ of an operator A in $\mathbf{C}^n$ by the generalized integral formula of Cauchy

$$f(A) = -\frac{1}{2\pi} \int_C f(\lambda) R_\lambda(A) d\lambda, \tag{1.1}$$

where C is a closed smooth contour surrounding $\sigma(A)$. If an analytic function $f(\lambda)$ is represented by the series

$$f(\lambda) = \sum_{k=0}^{\infty} c_k \lambda^k,$$

and the series

$$\sum_{k=0}^{\infty} c_k A^k$$

converges in the norm of the space $\mathbf{C}^n$, then (1.1) gives

$$f(A) = \sum_{k=0}^{\infty} c_k A^k.$$

Example 1.1.1 *Let A be a diagonal matrix:*

$$A = \begin{pmatrix} a_1 & 0 & \cdots & 0 \\ 0 & a_2 & \cdots & 0 \\ \cdot & \cdots & \cdot & \cdot \\ 0 & \cdots & 0 & a_n \end{pmatrix}.$$

Then one can show that

$$f(A) = \begin{pmatrix} f(a_1) & 0 & \cdots & 0 \\ 0 & f(a_2) & \cdots & 0 \\ \cdot & \cdots & \cdot & \cdot \\ 0 & \cdots & 0 & f(a_n) \end{pmatrix}.$$

Example 1.1.2 *If a matrix J is a Jordan block:*

$$J = \begin{pmatrix} \lambda_0 & 1 & 0 & \ldots & 0 \\ 0 & \lambda_0 & 1 & \ldots & 0 \\ \cdot & \cdot & \cdot & \ldots & \cdot \\ \cdot & \cdot & \cdot & \ldots & \cdot \\ \cdot & \cdot & \cdot & \ldots & \cdot \\ 0 & 0 & \ldots & \lambda_0 & 1 \\ 0 & 0 & \ldots & 0 & \lambda_0 \end{pmatrix},$$

then

$$f(J) = \begin{pmatrix} f(\lambda_0) & \frac{f'(\lambda_0)}{1!} & \ldots & \frac{f^{(n-1)}(\lambda_0)}{(n-1)!} \\ 0 & f(\lambda_0) & \ldots & \\ \cdot & \cdot & \ldots & \cdot \\ \cdot & \cdot & \ldots & \cdot \\ \cdot & \cdot & \ldots & \cdot \\ 0 & \ldots & f(\lambda_0) & \frac{f'(\lambda_0)}{1!} \\ 0 & \ldots & 0 & f(\lambda_0) \end{pmatrix}.$$

For more details about matrix-valued functions we refer the reader to (Gantmaher, 1967).

The following quantity plays a key role in the sequel

$$g(A) = (N^2(A) - \sum_{k=1}^{n} |\lambda_k(A)|^2)^{1/2},$$

where $N(A)$ is the Frobenius (Hilbert-Schmidt) norm of A, i.e.

$$N^2(A) = Trace(AA^*).$$

If A has in some orthonormed basis the entries a_{jk}, i.e.

$$A = (a_{jk})_{j,k=1}^{n},$$

then

$$N(A) = \sqrt{\sum_{j,k=1}^{n} |a_{jk}|^2}$$

and it does not depend on an orthonormed basis.

Since

$$\sum_{k=1}^{n} |\lambda_k(A)|^2 \geq |TrA^2|,$$

we get

$$g^2(A) \le N^2(A) - |TrA^2|.$$

If A is a normal matrix, i.e. if $AA^* = A^*A$, then $g(A) = 0$ (see Corollary 1.3.7 below).

We will also prove the inequality

$$g^2(A) \le \frac{1}{2}N^2(A^* - A) \tag{1.2}$$

(see Corollary 1.3.7 below).

It is easy to see that

$$N(A) \le \sqrt{n}\|A\|.$$

Hence

$$g(A) \le [n\|A\|^2 - \sum_{k=1}^{n} |\lambda_k(A)|^2]^{1/2} \le \sqrt{n}\|A\|.$$

Example 1.1.3 *Let $n = 2$ and a matrix A have real entries, only. That is,*

$$A = \begin{pmatrix} a_{11} & a_{12} \\ a_{21} & a_{22} \end{pmatrix}$$

and a_{jk}, $j, k = 1, 2$ are real numbers.

First consider the case of *nonreal eigenvalues*: $\lambda_2(A) = \overline{\lambda}_1(A)$. It can be written

$$det(A) = \lambda_1(A)\overline{\lambda}_1(A) = |\lambda_1(A)|^2$$

and

$$|\lambda_1(A)|^2 + |\lambda_2(A)|^2 = 2|\lambda_1(A)|^2 = 2det(A) = 2[a_{11}a_{22} - a_{21}a_{12}].$$

Thus,

$$g^2(A) = N^2(A) - |\lambda_1(A)|^2 - |\lambda_2(A)|^2 =$$
$$a_{11}^2 + a_{12}^2 + a_{21}^2 + a_{22}^2 - 2[a_{11}a_{22} - a_{21}a_{12}].$$

Hence,

$$g(A) = \sqrt{(a_{11} - a_{22})^2 + (a_{21} + a_{12})^2}.$$

Now let $n = 2$ and a matrix A have real entries again, but the *eigenvalues of A are real*. We have

$$|\lambda_1(A)|^2 + |\lambda_2(A)|^2 = \lambda_1^2(A) + \lambda_2^2(A) = Trace A^2.$$

Obviously,

$$A^2 = \begin{pmatrix} a_{11}^2 + a_{12}a_{21} & a_{11}a_{12} + a_{12}a_{22} \\ a_{21}a_{11} + a_{21}a_{22} & a_{22}^2 + a_{21}a_{12} \end{pmatrix}.$$

We thus get

$$|\lambda_1(A)|^2 + |\lambda_2(A)|^2 = a_{11}^2 + 2a_{12}a_{21} + a_{22}^2.$$

Consequently,

$$g^2(A) = N^2(A) - |\lambda_1(A)|^2 - |\lambda_2(A)|^2 =$$

$$a_{11}^2 + a_{12}^2 + a_{21}^2 + a_{22}^2 - (a_{11}^2 + 2a_{12}a_{21} + a_{22}^2).$$

Hence,

$$g(A) = \sqrt{|a_{12} - a_{21}|}.$$

Example 1.1.4 *Let A be an upper-triangular matrix*

$$A = \begin{pmatrix} a_{11} & a_{12} & \cdots & a_{1n} \\ 0 & a_{21} & \cdots & a_{2n} \\ \cdot & \cdots & \cdot & \cdot \\ 0 & \cdots & 0 & a_{nn} \end{pmatrix}.$$

Then

$$N(A) = \sqrt{\sum_{k=1}^{n} \sum_{j=1}^{k-1} |a_{jk}|^2},$$

since eigenvalues of a triangular matrix are its diagonal elements.

Example 1.1.5 *Consider the matrix A defined by*

$$A = \begin{pmatrix} -a_1 & \cdots & -a_{n-1} & -a_n \\ 1 & \cdots & 0 & 0 \\ \cdot & \cdots & \cdot & \cdot \\ 0 & \cdots & 1 & 0 \end{pmatrix}.$$

with real positive numbers a_k. Such matrices play a key role in the theory of scalar ordinary differential equations. Take into account that

$$A^2 = \begin{pmatrix} a_1^2 - a_2 & \cdots & a_1 a_{n-1} & a_1 a_n \\ -a_1 & \cdots & -a_{n-1} & -a_n \\ 0 & \cdots & 0 & 0 \\ \cdot & \cdots & \cdot & \cdot \\ 0 & \cdots & 0 & 0 \end{pmatrix}.$$

We thus obtain,

$$Trace\, A^2 = a_1^2 - 2a_2.$$

Therefore

$$g^2(A) \leq N^2(A) - |Trace\, A^2| \leq$$

$$\sum_{k=1}^{n} a_k^2 + n - 1 - a_1^2 + 2a_2.$$

Hence

$$g(A) \leq \sqrt{\sum_{k=3}^{n} a_k^2 + n - 2 + (1 + a_2)^2}.$$

Introduce for $n > 1$ the numbers

$$\gamma_{n,p} = \sqrt{\frac{C_{n-1}^p}{(n-1)^p}}$$

for $p = 1, ..., n - 1$, and $\gamma_{n,0} = 1$. Here

$$C_{n-1}^p = \frac{(n-1)!}{(n-p-1)!p!}$$

are binomial coefficients. Evidently, for $n > 2$

$$\gamma_{n,p}^2 = \frac{(n-2)(n-3)\ldots(n-p)}{(n-1)^{p-1}p!} \leq \frac{1}{p!}. \tag{1.3}$$

1.2   Estimations for matrix-valued functions

Theorem 1.2.1 *Let A be a linear operator in a complex Euclidean space C^n, and let f be a function regular in a neighborhood of the closed convex hull $co(A)$ of the spectrum of A. Then the inequality*

$$\|f(A)\| \leq \sum_{k=0}^{n-1} \sup_{\lambda \in co(A)} |f^{(k)}(\lambda)| g^k(A) \frac{\gamma_{n,k}}{k!} \tag{2.1}$$

is true.

The proofs of Theorems 1.2.1 and 1.2.2 are given in Section 2.4.

Theorem 1.2.1 is precise. The inequality (2.1) becomes the equality

$$\|f(A)\| = \sup_{\lambda \in \sigma(A)} |f(\lambda)|$$

if A is a normal matrix and

$$\sup_{\lambda \in \sigma(A)} |f(\lambda)| = \sup_{\lambda \in co(A)} |f(\lambda)|,$$

since $g(A) = 0$ in this case.

Theorem 1.2.2 *Let A be a linear operator in a complex Euclidean space C^n. Then the following estimate is true:*

$$\|R_\lambda(A)\| \leq \sum_{k=0}^{n-1} g^k(A) \frac{\gamma_{n,k}}{\rho^{k+1}(A,\lambda)} \quad \text{for all regular } \lambda, \tag{2.2}$$

where $\rho(A, \lambda)$ is the distance between the spectrum $\sigma(A)$ of A and a complex point λ.

Theorem 1.2.2 is precise. The inequality (2.2) becomes the equality

$$\|R_\lambda(A)\| = \rho^{-1}(A, \lambda)$$

if A is a normal matrix since $g(A) = 0$ in this case.

Notice that according to (1.3) we have the following results.

Corollary 1.2.3 *Under the conditions of Theorem 1.2.1 the inequality*

$$\|f(A)\| \leq \sum_{k=0}^{n-1} \sup_{\lambda \in co(A)} |f^{(k)}(\lambda)| \frac{g^k(A)}{(k!)^{3/2}} \tag{2.3}$$

is valid.

and

Corollary 1.2.4 *Under the conditions of Theorem 1.2.2 the inequality*

$$\|R_\lambda(A)\| \leq \sum_{k=0}^{n-1} \frac{g^k(A)}{\rho^{k+1}(A,\lambda)\sqrt{k!}} \text{ for all regular } \lambda \tag{2.4}$$

is true.

Example 1.2.5 *Let A be a linear operator in $\mathbf{C}^n$, then the following inequality:*

$$\|A^m\| \leq \sum_{k=0}^{m_1} \frac{m! d^{m-k}(A) g^k(A) \gamma_{n,k}}{(m-k)!k!}$$

holds for every integer m. Here

$$d(A) = \max_k |\lambda_k(A)|$$

is the spectral radius of A, and

$$m_1 = \min\{m, n-1\}.$$

Example 1.2.6 *Let A be a linear operator in $\mathbf{C}^n$. Consider the function $f(\lambda) = \exp(t\lambda)$ with a real positive number t.*

Theorem 1.2.1 gives in this case the following estimate:

$$\|exp(At)\| \leq e^{\alpha(A)t} \sum_{k=0}^{n-1} g^k(A) t^k \frac{\gamma_{n,k}}{k!}, \text{ for all } t \geq 0, \tag{2.5}$$

where

$$\alpha(A) = \max_{k=1,\ldots,n} \Re\lambda_k(A).$$

According to (1.3)

$$\|exp(At)\| \leq e^{\alpha(A)t} \sum_{k=0}^{n-1} \frac{g^k(A) t^k}{(k!)^{3/2}}, \text{ for all } t \geq 0. \tag{2.6}$$

Example 1.2.7 *Let A be an invertible n × n-matrix.*

Take the function $f(\lambda) = \lambda^{-1}$. Then by Theorem 1.2.2 we obtain

$$\|A^{-1}\| \leq \sum_{k=0}^{n-1} g^k(A) \frac{\gamma_{n,k}}{\rho_0^{k+1}(A)},$$

where $\rho_0(A)$ is the smallest modulus of the eigenvalues of A:

$$\rho_0(A) = \inf_{k=1,\ldots,n} |\lambda_k(A)|.$$

1.3 Preliminaries

1.3.1 Powers of a nilpotent matrix

Let $\mathbf{B}(\mathbf{C}^n)$ be the set of all linear operators in $\mathbf{C}^n$.

We recall that an operator $P \in \mathbf{B}(\mathbf{C}^n)$ is *a projector* if

$$P^2 = P.$$

$P \in \mathbf{B}(\mathbf{C}^n)$ is *an orthogonal projector (an orthoprojector)* if it is a projector and

$$P = P^*.$$

A subspace $M \subset \mathbf{C}^n$ is *an invariant subspace* of A if the relation $h \in M$ implies $Ah \in M$. If P is a projector onto an invariant subspace of A, then

$$PAP = AP.$$

By Schur's theorem (Marcus and Minc (1964), Section I.4.10.2) a linear operator $A \in \mathbf{B}(\mathbf{C}^n)$ can be decomposed as

$$A = D + V \tag{3.1}$$

with a normal (diagonal) operator D and a nilpotent (upper-triangular) operator V. In addition, all the invariant subspaces D and V are common.

We will call (3.1) *the triangular representation of A*, D the *diagonal part of A* and V the *nilpotent part of A*, respectively

Definition 1.3.1 *Let V be the nilpotent part of an operator $A \in B(\mathbb{C}^n)$, then the smallest integer ν such that $V^\nu = 0$ will be called the nilpotent index of A and we will write $\nu = ni(A)$.*

Evidently $ni(A) \le n$ and $ni(A) = ni(V)$.

Let $V \in B(\mathbb{C}^n)$ be a nilpotent operator and let $P_0, P_1, \ldots, P_n$ be orthogonal projectors onto its invariant subspaces: $0 \equiv P_0$, $P_n = I$, and

$$Range(P_0) \subset Range(P_1) \subset \ldots \subset Range(P_n).$$

Definition 1.3.2 *Let V be a nilpotent operator in $\mathbb{C}^n$. Then the greatest integer r such that*

$$P_{k-r} V P_k = V P_k \quad \text{for all} \quad k = r + 1, \ldots, n$$

will be called the dimension of the zero-upperdiagonal of V and we will write $r = dzud(V)$.

Lemma 1.3.3 *Let $V \in B(\mathbb{C}^n)$ be a nilpotent operator, then*

$$r(\nu - 1) \le n - 1.$$

Here $r = dzud(V)$, $\nu = ni(V)$.

Proof: We have by Definition 1.3.2

$$V^2 P_k = V P_{k-r} V P_k V P_k = P_{k-2r} V P_{k-r} V P_k = P_{k-2r} V^2 P_k.$$

I.e.

$$dzud(V^2) \ge 2dzud(V).$$

Similarly,

$$dzud(V^m) \ge m \, dzud(V) \quad \text{for all } m < n. \tag{3.2}$$

Since $V^\nu = 0$, then

$$dzud(V^{\nu-1}) \le n - 1.$$

By (3.2) $n - 1 \ge (\nu - 1)r$. As claimed. $\square$

Lemma 1.3.4 *Let $Q, V \in B(\mathbf{C}^n)$ and let V be a nilpotent operator. Suppose that all invariant subspaces of V and of Q are common. Then VQ and QV are nilpotent operators. Moreover,*

$$dzud(QV) \geq dzud(V) \ and \ dzud(VQ) \geq dzud(V).$$

Proof: Let $P_1, P_2, \ldots, P_n$ be projectors onto invariant subspaces of both V and Q. By Definition 1.3.2

$$QVP_k = QP_{k-r}VP_k = P_{k-r}QVP_k$$

as long as $k = r+1, \ldots, n$. Here $r = dzud(V)$. That is,

$$dzud(QV) \geq dzud(V).$$

Similarly, we will prove the other inequality. □

We introduce the following numbers

$$\gamma_{n,p}^{(r)} = \left(\sum_{1 \leq k_1 \leq k_2 - r \leq \ldots \leq k_p - r(p-1) \leq n - pr} 1 \right)^{1/2} (n-r)^{-p/2}$$

for any integers p and r satisfying the condition $pr < n$ for $p \geq 1$, and $\gamma_{n,0}^{(r)} = 1$. It is simple to show that

$$\gamma_{n,p}^{(r)} \leq \gamma_{n,p}^{(r-1)} \ \text{for all} \ r > 1$$

and

$$\gamma_{n,p}^{(1)} = \left(\sum_{1 \leq k_1 < k_2 < \ldots < k_p \leq n-1} \right)^{1/2} (n-1)^{-p/2} = [C_{n-1}^p (n-1)^{-p}]^{1/2} = \gamma_{n,p}.$$

That is,

$$\gamma_{n,p}^{(r)} \leq \gamma_{n,p}^{(1)} = \gamma_{n,p}. \tag{3.3}$$

Lemma 1.3.5 *Let $V \in B(\mathbf{C}^n)$ be a nilpotent operator and $r = dzud(V)$, then*

$$\|V^p\| \leq \gamma_{n,p}^{(r)} N^p(V)$$

for each $p = 1, \ldots, ni(V) - 1$.

Proof: Denote

$$\|x\|_m = (\sum_{k=m}^{n} |x_k|^2)^{1/2} \text{ for } m < n,$$

where x_k are coordinates of a vector x. Let a_{jk} be entries of the upper-triangular matrix V. We can write

$$\|Vx\|_m^2 = \sum_{j=m}^{n-r} | \sum_{k=j+r}^{n} a_{jk} x_k|^2, \text{ for all } m < n - r$$

taking into account that $dzud(V) = r$. Now, we have (by Schwartz's inequality) the relation

$$\|Vx\|_m^2 \le \sum_{j=m}^{n-r} h_j \|x\|_{j+r}^2, \tag{3.4}$$

where

$$h_j = \sum_{k=j+r}^{n} |a_{jk}|^2.$$

Let $a_{jk}^{(2)}$ be entries of V^2. Due to (3.2)

$$a_{jk}^{(2)} = 0 , j > k - 2r , k = 2r, ..., n.$$

Taking into account (3.4) we obtain

$$\|V^2 x\|_m^2 = \sum_{j=m}^{n-2r} | \sum_{k=2r+j}^{n} a_{jk}^{(2)} x|^2 =$$

$$\sum_{j=m}^{n-2r} | \sum_{k=j+r}^{n} a_{jk} (Vx)_k|^2 \le \sum_{j=m}^{n-2r} h_j \|Vx\|_{j+r}^2.$$

Here $(Vx)_k$ are coordinates of Vx. From this it follows that

$$\|V^2 x\|_m^2 \le \sum_{j=m}^{n-2r} h_j \sum_{k=j+r}^{n-r} h_k \|x\|_{k+r}^2 =$$

$$\sum_{m \le j \le k-r \le n-2r} h_j h_k \|x\|_{k+r}^2.$$

Repeating these arguments, we arrive at the inequality

$$\|V^p x\|_1^2 \leq \sum_{1 \leq k_1 \leq k_2 - r \leq \ldots \leq k_p - r(p-1) \leq n - pr} h_{k_1} \ldots h_{k_p} \qquad (3.5)$$

when $\|x\| = \|x\|_1 = 1$.

Put

$$Q(h_1, ..., h_n) = \Big(\sum_{1 \leq k_1 \leq k_2 - r \leq \ldots \leq k_p - r(p-1) \leq n - pr} h_{k_1} h_{k_2} \ldots h_{k_p} \Big) \Big(\sum_{k=1}^{n-r} h_k \Big)^{-p}.$$

Since $Q(h_1, ..., h_n)$ is a symmetric function of $h_1, h_2, \ldots, h_n$, it is easy to show that $Q(h_1, ..., h_n)$ has a maximum when $h_1 = h_2 = \ldots = h_n$. Hence

$$\max_{h_1, h_2, \ldots, h_n} Q(h_1, ..., h_n) = [\gamma_{m,p}^{(r)}]^2.$$

Since

$$\sum_{k=1}^{n-r} h_j \leq N^2(V),$$

the required result follows from (3.5).□

1.3.2 Some properties of $g(A)$

Again, let V be the nilpotent part of a matrix A, P_j $(j = 0, 1, ..., n)$ are projectors onto invariant subspaces of A,

Lemma 1.3.6 *The equalities*

$$N^2(VP_j) = N^2(AP_j) - \sum_{k=1}^{j} |\lambda_k|^2 =$$

$$2N^2(P_j A_I P_j) - 2\sum_{k=1}^{j} |\Im \lambda_k|^2$$

are true for any $A \in B(\mathbf{C}^n)$ and $j = 1, \ldots, n$ where $\lambda_k = \lambda_k(AP_j) = \lambda_k(A), k = 1, \ldots, j$ are eigenvalues of AP_j taking into account their multiplicities.

Proof: Due to Lemma 1.3.4 both matrices V^*D and D^*V are nilpotent. Therefore,

$$Trace(P_j D^* V P_j) = 0 \text{ and } Trace(P_j V^* D P_j) = 0. \qquad (3.6)$$

It is easy to see that

$$Trace(D^* D P_j) = \sum_{k=1}^{j} |\lambda_k(A)|^2 \text{ for } j = 1, \ldots, n. \qquad (3.7)$$

Due to (3.6) and (3.7) we can write down

$$N^2(A P_j) = Trace\{P_j (D + V)^* (V + D) P_j\} =$$

$$Trace\{P_j V^* V P_j + D^* D P_j\} =$$

$$N^2(V P_j) + \sum_{k=1}^{j} |\lambda_k(A)|^2, \qquad (3.8)$$

and the left part of the required relation is proved.

On the other hand,

$$-4(P_j A_I P_j)^2 = P_j(A - A^*)P_j(A - A^*)P_j =$$

$$P_j A P_j A P_j - P_j A P_j A^* P_j - P_j A^* P_j A P_j + P_j A^* P_j A^* P_j.$$

From this it follows that

$$-4(P_j A_I P_j)^2 = P_j A^2 P_j - P_j A A^* P_j - P_j A^* A P_j +$$

$$P_j A^* A^* P_j = P_j (A - A^*)^2 P_j.$$

Now (3.1) implies

$$4N^2(P_j A_I P_j) = -Trace(P_j(A - A^*)^2 P_j) =$$

$$-Trace(P_j(V^* + D^* - V - D)^2 P_j).$$

Due to (3.6) we have

$$4N^2(P_j A_I P_j) = -Trace(P_j(V - V^*)^2 P_j) - Trace(P_j(D - D^*)^2 P_j).$$

Set

$$V_I = (V - V^*)/2i$$

and

$$D_I = (D - D^*)/2i.$$

Thus we have obtained the equality

$$N^2(P_j A_I P_j) = N^2(P_j V_I P_j) + N^2(D_I P_j). \tag{3.9}$$

It is not hard to see that

$$N^2(P_j V_I P_j) = \frac{1}{2} \sum_{m=1}^{j} \sum_{k=1}^{m-1} |a_{km}|^2 = \frac{1}{2} N^2(V P_j).$$

From this and from (3.9) it follows that

$$N^2(V P_j) = 2N^2(P_j A_I P_j) - 2N^2(D_I P_j).$$

Comparing this and (3.8) we arrive at the result since

$$N^2(D_I P_j) = \sum_{k=1}^{j} |\Im \lambda_k|^2. \ \square$$

Corollary 1.3.7 *For any* $A \in B(\mathbb{C}^n)$ *the relation*

$$N^2(V) = g^2(A) \equiv N^2(A) - \sum_{k=1}^{n} |\lambda_k|^2 =$$

$$2N^2(A_I) - 2\sum_{k=1}^{n} |\Im \lambda_k|^2 \tag{3.10}$$

is true.

Indeed, this result follows from Lemma 1.3.6 when $j = n$.

Take into account that the norms of nilpotent parts of the matrices A and $Ae^{it} + zI$ with a real t and a complex number z coincide. We obtain the following

Corollary 1.3.8 *For any* $A \in B(\mathbf{C}^n)$, *a real number* t, *and a complex one* z, *the relation*

$$g(Ae^{it} + zI) = g(A) \tag{3.11}$$

holds.

Corollary 1.3.9 *For any commuting* $A, B \in B(\mathbf{C}^n)$ *the relation*

$$g(A + B) \leq g(A) + g(B)$$

holds.

In fact, A and B have the same Schur basis. This clearly forces

$$V_{A+B} = V_A + V_B,$$

where V_{A+B}, V_A and V_B are the nilpotent parts of $A + B, A$ and B, respectively. Due to Corollary 1.3.7 the relations

$$g(A) = N(V_A), \ g(B) = N(V_B), \ g(A + B) = N(V_{A+B})$$

are true. Now the property of the norm implies the result.

Lemma 1.3.10 *For any* $A \in B(\mathbf{C}^n)$ *and real numbers* t, τ *the relation*

$$N^2(Ae^{it} - A^*e^{-it}) - \sum_{k=1}^{n} |e^{it}\lambda_k - e^{-it}\overline{\lambda}_k|^2 =$$

$$N^2(Ae^{i\tau} - A^*e^{-i\tau}) - \sum_{k=1}^{n} |e^{i\tau}\lambda_k - e^{-i\tau}\overline{\lambda}_k|^2$$

holds.

Proof: Indeed , put

$$Q = Ae^{it} \text{ and } T = Ae^{i\tau}.$$

Due to Corollary 1.3.8 $g(T) = g(Q)$. Corollary 1.3.7 implies

$$g^2(T) = 2N^2(T_I) - 2\sum_{k=1}^{n} |\Im\lambda_k(T)|^2$$

and

$$g^2(Q) = 2N^2(Q_I) - 2\sum_{k=1}^{n} |\Im\lambda_k(Q)|^2,$$

where Q_I and T_I are the imaginary components of the matrices Q and T, respectively. This forces the equality

$$N^2(T_I) - \sum_{k=1}^{n} |\Im\lambda_k(T)|^2 = N^2(Q_I) - \sum_{k=1}^{n} |\Im\lambda_k(Q)|^2$$

which is our claim assertion, because

$$\lambda_k(T) = e^t\lambda_k(A) \text{ and } \lambda_k(Q) = e^\tau\lambda_k(A).\Box$$

1.3.3   Auxiliary inequalities

Lemma 1.3.11 *Let A be a linear operator in $\mathbf{C}^n$ and $\{d_k\}$ be an orthonormed basis in $\mathbf{C}^n$. Then for each $x \in \mathbf{C}^n$*

$$\|Ax\| \le \| \, |A||x| \, \|$$

where $|A|$ is the matrix whose elements with respect to $\{d_k\}$ are the moduli of the elements of A with respect to $\{d_k\}$, and $|x|$ is the vector whose coordinates are the moduli of the coordinates of the vector x in the basis $\{d_k\}$.

That is,

$$|x| = \sum_{k=1}^{n} |x_k|d_k \text{ with } x_k = (x, d_k).$$

Proof: The proof is the evident application of the equality

$$\|Ax\|^2 = \sum_{j=1}^{n} |(Ax)_j|^2,$$

where $(.)_j$ means the j-th coordinate. $\Box$

Lemma 1.3.12 *Let $\{d_k\}$ be an orthonormed basis in $\mathbf{C}^n$, $A_1, \ldots, A_j$ be operators in $\mathbf{C}^n$, and $\phi(k_1, \ldots, k_{j+1})$ be a scalar-valued function of $j + 1$ natural arguments. Define orthoprojectors $Q(k)$ by*

$$Q(k)h = (h, d_k)d_k, \ k = 1, \ldots, n,$$

and set

$$T = \sum_{1 \le k_1 \le \dots \le k_{j+1} \le n} \phi(k_1, \dots, k_{j+1}) Q(k_1) A_1 Q(k_2) \dots A_j Q(k_{j+1}).$$

Then for any $h \in \mathbf{C}^n$ the following inequality is true

$$|Th| \le m_0 |A_1| |A_2| \dots |A_j| |h|,$$

where

$$m_0 = \max_{1 \le k_1 \le \dots \le k_{j+1} \le n} |\phi(k_1, \dots, k_{j+1})|,$$

and $|A_k|$ are operators whose matrix elements are moduli of matrix elements of A_k with respect to the basis $\{d_k\}$.

Proof: For any coordinate $(Th)_l$ of the vector Th we have

$$(Th)_l = Q(l)Th = \sum_{1 \le k_2 \le \dots \le k_{j+1} \le n} \phi(l, \dots, k_{j+1}) a^{(1)}_{l k_2} \dots a^{(j)}_{k_j k_{j+1}} h_{k_{j+1}},$$

where

$$a^{(i)}_{k_i, k_{i+1}} = Q(k_i) A_i Q(k_{i+1})$$

are elements of A_i. From this and from Lemma 1.3.11 the result trivially follows. $\square$

Consider the integral

$$J \equiv \frac{1}{2\pi i} \int_C \frac{f(\lambda) d\lambda}{(\lambda - a_1) \dots (\lambda - a_{n+1})},$$

where C is a smooth closed contour surrounding numbers $a_1, \dots, a_{n+1}$, and $f(\lambda)$ is a function which is analytic on the closed convex hull *co* of numbers $a_1, \dots, a_{n+1}$.

Lemma 1.3.13 *The estimate*

$$|J| \le \max_{\lambda \in co} \frac{|f^{(n)}(\lambda)|}{n!}$$

is true.

Proof: By the residue theorem it follows that

$$J = \sum_{k=1}^{n+1} \frac{f(x_k)}{\prod_{s=1, s \neq k}^{n+1}(x_k - x_s)}.$$

But due to the well-known formula (48) from Gelfond (1967) the expression in the right part is the k-th order divided difference of f in the points $a_1, \ldots, a_{k+1}$. Now the well known estimate for the divided difference (49) from Gelfond (1967) gives the result. $\square$

1.4   Proofs of Theorems 1.2.1 and 1.2.2

Let $|V|$ be the operator whose matrix elements in the orthonormed basis of the triangular representation (the Schur basis) of the operator A are the moduli of the matrix elements of the nilpotent part V of A with respect to this basis (i.e. the elements of $|V|$ are the moduli of the elements of V, when V has an upper-triangular form). This means that

$$|V| = \sum_{k=1}^{n} \sum_{j=1}^{k-1} |a_{jk}|(., e_k)e_j,$$

where $\{e_k\}$ is the Schur basis and

$$a_{jk} = (Ae_k, e_j) \text{ for } j \leq k; \ a_{jk} = 0 \text{ for } j > k.$$

Lemma 1.4.1 *Under the hypothesis of Theorem 1.2.1 the following estimate*

$$\|f(A)\| \leq \sum_{k=0}^{n-1} \sup_{\lambda \in co(A)} |f(k)(\lambda)| \frac{\| |V|^k \|}{k!}$$

is true.

Proof: It is not hard to see that the representation (3.1) implies the equality

$$R_\lambda(A) \equiv (A_\lambda - I)^{-1} = (D + V - \lambda I)^{-1} =$$

$$(I + R_\lambda(D)V)^{-1} R_\lambda(D) \tag{4.1}$$

for all regular λ. According to Lemma 1.3.4 $R_\lambda(D)V$ is a nilpotent operator because V and $R_\lambda(D)$ have common invariant subspaces. Hence

$$(R_\lambda(D)V)^n = 0.$$

Therefore

$$(I + R_\lambda(D)V)^{-1} = \sum_{k=0}^{n-1}(R_\lambda(D)V)^k(-1)^k. \qquad (4.2)$$

Due to (1.1)

$$f(A) = \sum_{k=0}^{n-1} C_k, \qquad (4.3)$$

where

$$C_k = (-1)^{k+1}\frac{1}{2\pi i}\int_C f(\lambda)(R_\lambda(D)V)^k R_\lambda(D)d\lambda.$$

Since D is a diagonal matrix with respect to the Schur basis, then

$$R_\lambda(D) = \sum_{j=1}^{n}\frac{Q_j}{a_{jj} - \lambda},$$

where

$$Q_k = (.,e_k)e_k = \Delta P_k.$$

Since $Q_j V Q_k = 0$ for $j \geq k$, from (4.3) it follows that

$$C_k = \sum_{j_1=1}^{n} Q_{j_1} V \sum_{j_2=1}^{n} Q_{j_2} V \ldots V \sum_{j_k=1}^{n} Q_{j_k} I_{j_1 \ldots j_{k+1}}.$$

Here

$$I_{j_1 \ldots j_{k+1}} = \frac{(-1)^{k+1}}{2\pi i}\int_C \frac{f(\lambda)d\lambda}{(a_{j_1 j_1} - \lambda)\ldots(a_{j_{k+1} j_{k+1}} - \lambda)}.$$

Lemma 1.3.12 gives the estimate

$$\|C_k\| \leq \max_{1 \leq j_1 \leq \ldots \leq j_{k+1} \leq n} |I_{j_1 \ldots j_{k+1}}| \| |V|^k \|.$$

Due to Lemma 1.3.11

$$I_{j_1 \ldots j_{k+1}} \leq \sup_{\lambda \in co(A)} \frac{|f^{(k)}(\lambda)|}{k!}.$$

This inequality and (4.3) imply the result.$\square$

Denote by $[x]$ the integral part of a positive number x.

Lemma 1.4.2 *Let A be a linear operator in $\mathbf{C}^n$, and let f be a function regular in a neighborhood of the closed convex hull $co(A)$ of the spectrum of A. Then the following inequality holds*

$$\|f(A)\| \le \sum_{k=0}^{n_1} \sup_{\lambda \in co(A)} |f^{(k)}(\lambda)| \frac{g^k(A)\gamma_{n,k}^{(r)}}{k!},$$

where

$$n_1 = [\frac{n-1}{r}]$$

and $r = dzud(V)$.

Proof: Lemma 1.3.5 implies

$$\| \, |V|^k \, \| \le \gamma_{n,k}^{(r)} N^k(|V|) \, , k = 1, ..., n.$$

It is easy to see that

$$dzud(|V|) = dzud(V) = r \text{ and } N(|V|) = N(V).$$

Due to Corollary 1.3.7 we have

$$N(V) = g(A).$$

According to Lemma 1.3.3 the inequality

$$ni(|V|) \le 1 + [(n-1)/r] = 1 + n_1$$

holds. This means that $|V|^{1+n_1} = 0$. Now Lemma 1.4.1 implies the result.□

Proof of Theorem 1.2.1: The assertion of Theorem 1.2.1 follows immediately from Lemma 1.4.2, because $r \ge 1$ for any nilpotent matrix. Thus, $n_1 \le n - 1$. For the completion of the proof it remains to recall inequality (3.3).□

Lemma 1.4.3 *Let A be a linear operator in a complex Euclidean space $\mathbf{C}^n$. Then the estimate*

$$\|R_\lambda(A)\| \le \sum_{k=0}^{n_1} g^k(A)\gamma_{n,k}^{(r)}\rho^{-k-1}(A,\lambda) \text{ for all regular } \lambda,$$

is true.

Here
$$n_1 = [\frac{n-1}{r}]$$

and $r = dzud(V)$, again.

Proof: We use the relation (4.1). According to Lemma 1.3.4

$$dzud(R_\lambda(D)V) \geq dzud(V) = r$$

because V and $R_\lambda(D)$ have common invariant subspaces. By Lemma 1.3.3

$$ni(R_\lambda(D)V) \leq 1 + [(n-1)/r] = 1 + n_1.$$

Hence,
$$(R_\lambda(D)V)^{n_1+1} = 0.$$

Therefore,
$$(I + R_\lambda(D)V)^{-1} = \sum_{k=0}^{n_1}(R_\lambda(D)V)^k(-1)^k.$$

Lemma 1.3.5 gives the inequality

$$\|(R_\lambda(D)V)^k\| \leq N^k(R_\lambda(D)V)\gamma_{n,k}^{(r)}. \qquad (4.4)$$

Since D is a normal operator, we can write down

$$\|R_\lambda(D)\| = \rho^{-1}(D,\lambda). \qquad (4.5)$$

It is clear that
$$N(R_\lambda(D)V) \leq N(V)\,\|R_\lambda(D)\|. \qquad (4.6)$$

This and (4.2) imply the relations

$$\|R_\lambda(A)\| = \|(I + R_\lambda(D)V)^{-1}R_\lambda(D)\| \leq$$

$$\sum_{k=0}^{n_1}\|(R_\lambda(D)V)^k R_\lambda(D)\|. \qquad (4.7)$$

By (4.4) we have

$$\|R_\lambda(A)\| \leq \sum_{k=0}^{n_1}N^k(R_\lambda(D)V)\gamma_{n,k}^{(r)}\|R_\lambda(D)\|.$$

Thus,

$$\|R_\lambda(A)\| \le \sum_{k=0}^{n_1} N^k(V)\gamma_{n,k}^{(r)}\rho^{-k-1}(D,\lambda).$$

This relation and Corollary 1.3.7 prove the stated result since $\sigma(A) = \sigma(D)$.
□

Proof of Theorem 1.2.2: The assertion of Theorem 1.2.2 follows immediately from Lemma 1.4.3 because $r \ge 1$ for any nilpotent matrix, and, thus, $n_1 \le n - 1$. For the completion of the proof it remains to recall the inequality (3.3).□

Direct proofs of Corollaries 1.2.3 and 1.2.4 : One can prove Corollary 1.2.3 and Corollary 1.2.4 by simple Lemma 2.3.1 proved below instead of Lemma 1.3.5.

In fact, due to Lemma 2.3.1

$$\| |V|^k \| \le \frac{N^k(|V|)}{\sqrt{k!}} \ , k = 1, 2, ...,$$

where $|V|$ is the same that in Lemma 1.4.1. Take into account that $N(|V|) = N(V)$. Now Corollary 1.3.7 gives

$$\| |V|^k \| \le \frac{g^k(A)}{\sqrt{k!}} \ , k = 1, 2, ...$$

This inequality and Lemma 1.4.1 prove Corollary 1.2.3 since $n_1 \le n - 1$.

Now we will prove Corollary 1.2.4. Indeed, Lemma 2.3.1 implies

$$\|(R_\lambda(D)V)^k\| \le \frac{N^k(R_\lambda(D)V)}{\sqrt{k!}} \ , k = 1, 2, ...,$$

since $R_\lambda(D)V$ is a nilpotent matrix. By (4.6) we obtain

$$\|(R_\lambda(D)V)^k\| \le \frac{N^k(V)\|R_\lambda(D)\|^k}{\sqrt{k!}} \ , k = 1, 2, ... \ .$$

Due to (4.5) and the equality $\sigma(A) = \sigma(D)$ we have

$$\|(R_\lambda(D)V)^k\| \le \frac{N^k(V)}{\rho^k(A,\lambda)\sqrt{k!}} \ , k = 1, 2, ... \ .$$

Now Corollary 1.3.7 gives

$$\|(R_\lambda(D)V)^k\| \le \frac{g^k(A)}{\rho^k(A,\lambda)\sqrt{k!}} \;, k = 1, 2, \dots \;.$$

It remains to take into account the relations (4.7) and $n_1 \le n - 1.\square$

1.5 Inequalities for eigenvalues of matrices

Let $B(\mathbf{C}^n)$ be the set of all linear operators in a Euclidean space $\mathbf{C}^n$, again. Let $\lambda_j = \lambda_j(A)$ and $s_j = s_j(A), (j = 1, \dots, n)$ be the eigenvalues taking into account the multiplicities of $A \in B(\mathbf{C}^n)$ and of $(AA^*)^{1/2}$, respectively. That is, $s_j(A)$ are the singular numbers of A.

By $b_j = b_j(A)$ we will denote the moduli of the eigenvalues of the matrix $A_I = (A - A^*)/2i$, i.e.

$$b_j = |\lambda_j(A_I)| \; ; j = 1, \dots, n.$$

In this section inequalities for λ_j, s_j and b_j are obtained.

Below we will assume that

$$s_1 \ge s_2 \ge \dots \ge s_n$$

and

$$b_1 \ge b_2 \ge \dots, \ge b_n.$$

The order of the eigenvalues λ_j is arbitrary.

Theorem 1.5.1 *For any $A \in B(\mathbf{C}^n)$, the following inequalities*

$$\sum_{k=1}^{j}(s_{n-k+1}^2 - |\lambda_k|^2) \le 2\sum_{k=1}^{j}(b_k^2 - |\Im\lambda_k|^2) \tag{5.1}$$

and

$$2\sum_{k=1}^{j}(b_{n-k+1}^2 - |\Im\lambda_k|^2) \le \sum_{k=1}^{j}(s_k^2 - |\lambda_k|^2) \; \text{for all } j = 1, \dots, n \tag{5.2}$$

are valid.

Besides, from our reasoning below it follows that

$$s_1 \ge |\lambda_j| \ge s_n \text{ and } b_1 \ge |\Im\lambda_j| \ge b_n \text{ for all } j = 1, \dots, n. \tag{5.3}$$

1.5.1   Proof of Theorem 1.5.1

We need the following result (see Marcus and Minc (1964), Section 4.1.5).
Let M be a non-negative Hermitian $n \times n$-matrix, and $\{g_k\}$ be an arbitrary
orthonormed basis in $\mathbf{C}^n$. Let also

$$\lambda_1(M) \geq \lambda_2(M) \geq \ldots \geq \lambda_n(M)$$

be eigenvalues of M accounting of their multiplicity. Then the relation

$$\sum_{k=1}^{j}(Mg_k, g_k) \geq \sum_{k=1}^{j}\lambda_{n-k+1}(M) \text{ for all } j = 1, ..., n \qquad (5.4)$$

is true.

Lemma 1.5.2 *Let the dimension of the null-space of the nilpotent part V
of a matrix A be equal to $r \geq 1$. Then*

$$\sum_{k=1}^{j} s_{n-k+1}^2 \leq \sum_{k=1}^{j}|\lambda_k|^2 \leq \sum_{k=1}^{j} s_k^2$$

and

$$\sum_{k=1}^{j} b_{n-k+1}^2 \leq \sum_{k=1}^{j}|\Im\lambda_k|^2 \leq \sum_{k=1}^{j} b_k^2$$

for $j = 1, ..., r$.

Proof: We can write down

$$VP_j = 0 \text{ for } j = 1, ..., r, \qquad (5.5)$$

where P_j are the projectors onto invariant subspaces of V. That is,

$$P_j = \sum_{k=1}^{j}(., e_k)e_k.$$

Here $\{e_k\}$ is the Schur basis, again. We can write

$$N^2(AP_j) = \sum_{k=1}^{j}(A^*Ae_k, e_k). \qquad (5.6)$$

According to (5.4) and (5.6), Lemma 1.3.6 gives the relations

$$\sum_{k=1}^{j} s_{n-k+1}^2 \leq N^2(VP_j) + \sum_{k=1}^{j}|\lambda_k|^2 = N^2(AP_j) \leq \sum_{k=1}^{j} s_k^2. \qquad (5.7)$$

Similarly, due to Lemma 1.3.6

$$\sum_{k=1}^{j} 2b_{n-k+1}^2 \leq N^2(VP_j) + 2\sum_{k=1}^{j}|\Im\lambda_k|^2 =$$

$$2N^2(P_j A_I P_j) \leq \sum_{k=1}^{j} 2b_k^2. \qquad (5.8)$$

Now the result follows from (5.5), (5.7) and (5.8).□

Proof of Theorem 1.5.1: The relations (5.7) and (5.8) imply:

$$\sum_{k=1}^{j}(s_{n-k+1}^2 - |\lambda_k|^2) \leq N^2(VP_j) \leq 2\sum_{k=1}^{j}(b_k^2 - |\Im\lambda_k|^2).$$

Therefore, (5.1) is proved. The inequality (5.2) is similarly proved. □

Proof of the inequality (5.3): Take into account that the dimension of the null-space of a nilpotent operator is greater than or equal to 1. Exploiting Lemma 1.5.2 we have

$$s_n \leq |\lambda_1| \leq s_1$$

and

$$b_n \leq |\Im\lambda_1| \leq b_1.$$

But the order of λ_j is arbitrary. Therefore we can take any λ_k as λ_1. This proves the result.□

1.6 Multiplicative structure of the resolvent of a matrix

We again denote by P_k ($k = 0,\ldots,n$) the projectors onto the invariant subspaces of $A \in B(\mathbf{C}^n)$ and use the triangular representation (3.1), V is the nilpotent part of A, and

$$D = \sum_{k=1}^{n} \lambda_k(A)\Delta P_k \qquad (6.1)$$

is the diagonal part. Here

$$\Delta P_k = P_k - P_{k-1}.$$

For $X_1, X_2, ..., X_n \in B(\mathbf{C}^n)$ denote

$$\overrightarrow{\prod_{1 \le k \le n}} X_k \equiv X_1 X_2 ... X_n.$$

That is, the arrow over the symbol of the product means that the indexes of the co-factors increase from left to right.

Theorem 1.6.1 *For any $A \in B(\mathbf{C}^n)$ and any regular λ the following representation*

$$\lambda R_\lambda(A) = - \overrightarrow{\prod_{1 \le k \le n}} (I + \frac{A\Delta P_k}{\lambda - \lambda_k(A)})$$

is true.

Proof: Denote $E_k = I - P_k$. Since

$$A = (E_k + P_k)A(E_k + P_k) \text{ for any } k = 1, ..., n$$

and $E_1 A P_1 = 0$, we get the relation

$$A = P_1 A E_1 + P_1 A P_1 + E_1 A E_1.$$

Take into account that $\Delta P_1 = P_1$ and

$$P_1 A P_1 = \lambda_1(A)\Delta P_1.$$

Then

$$A = \lambda_1(A)\Delta P_1 + P_1 A E_1 + E_1 A E_1 =$$

$$\lambda_1(A)\Delta P_1 + A E_1. \qquad (6.2)$$

Now, we check the equality

$$R_\lambda(A) = \Psi(\lambda), \tag{6.3}$$

where

$$\Psi(\lambda) \equiv \frac{\Delta P_1}{\lambda_1(A) - \lambda} - \frac{\Delta P_1}{\lambda_1(A) - \lambda} A E_1 R_\lambda(A) E_1 + E_1 R_\lambda(A) E_1.$$

In fact, multiplying this equality from the left by $A - I\lambda$ and taking into account the equality (6.2), we obtain the relation

$$(A - I\lambda)\Psi(\lambda) = \Delta P_1 - \Delta P_1 A E_1 R_\lambda(A) E_1 + (A - I\lambda) E_1 R_\lambda(A) E_1.$$

But $E_1 A E_1 = E_1 A$ and thus

$$E_1 R_\lambda(A) E_1 = E_1 R_\lambda(A).$$

I.e. we can write down

$$(A - I\lambda)\Psi(\lambda) = \Delta P_1 + (-\Delta P_1 A + A - I\lambda)E_1 R_\lambda(A) =$$
$$\Delta P_1 + E_1(A - I\lambda)R_\lambda(A) = \Delta P_1 + E_1 = I.$$

Similarly, we multiply (6.3) by $A - I\lambda$ from the right and take into account (6.2). This gives I. Therefore, (6.3) is correct.

Due to (6.3)

$$I - A R_\lambda(A) =$$
$$(I - (\lambda_1(A) - \lambda)^{-1} A \Delta P_1)(I - A E_1 R_\lambda(A) E_1). \tag{6.4}$$

Now we apply our reasoning above to the operator $A E_1$. We obtain the following expression which is similar to (6.4):

$$I - A E_1 R_\lambda(A) E_1 =$$
$$(I - (\lambda_2(A) - \lambda)^{-1} A \Delta P_2)(I - A E_2 R_\lambda(A) E_2).$$

For the operator $A P_k$ with any $k < n$ it similarly follows

$$I - A E_k R_\lambda(A) E_k =$$

$$(I - \frac{A\Delta P_{k+1}}{\lambda_{k+1}(A) - \lambda})(I - AE_{k+1}R_\lambda(A)E_{k+1}).$$

Substitute this in (6.4) as long as $k = 1, 2, ..., n-1$. It can be written

$$I - AR_\lambda(A) =$$

$$\overrightarrow{\prod_{1 \leq k \leq n-1}}(I + \frac{A\Delta P_k}{\lambda - \lambda_k(A)})(I - AE_{n-1}R_\lambda(A)E_{n-1}). \qquad (6.5)$$

It is clear that $E_{n-1} = \Delta P_n$. I.e.,

$$I - AE_{n-1}R_\lambda(A)E_{n-1} = I + \frac{A\Delta P_n}{\lambda - \lambda_n(A)}.$$

Now the identity

$$I - AR_\lambda(A) = -\lambda R_\lambda(A)$$

and (6.5) imply the result. □

Let A be a normal matrix. Then the representation

$$A = \sum_{k=1}^{n} \lambda_k(A)\Delta P_k$$

is true (see for instance Marcus and Minc (1964)). Hence,

$$A\Delta P_k = \lambda_k(A)\Delta P_k$$

Since

$$\Delta P_k \Delta P_j = 0 \text{ for } j \neq k,$$

Theorem 1.6.1 gives the equality

$$\lambda R_\lambda(A) = -\sum_{k=1}^{n}(I + (\lambda - \lambda_k(A))^{-1}\lambda_k(A)\Delta P_k).$$

But

$$I = \sum_{k=1}^{n} \Delta P_k.$$

The result is

$$\lambda R_\lambda(A) = -\sum_{k=1}^{n}[1 + (\lambda - \lambda_k(A))^{-1}\lambda_k(A)]\Delta P_k) =$$

$$-\sum_{k=1}^{n}\lambda\frac{\Delta P_k}{\lambda-\lambda_k(A)}.$$

Or

$$R_\lambda(A)=\sum_{k=1}^{n}\frac{\Delta P_k}{\lambda_k(A)-\lambda}.$$

We have obtained the well-known spectral representation for the resolvent of a normal matrix (see e.g. **Marcus and Minc (1964)**).

Thus, Theorem 1.6.1 generalizes the spectral representation for the resolvent of a normal matrix.

Let V be a nilpotent operator in $\mathbf{C}^n$ and P_k be the increasing chain of its invariant projectors. Then the relation

$$(I-V)^{-1}=\overrightarrow{\prod_{2\leq k\leq n}}(I+V\Delta P_k) \tag{6.6}$$

is valid.

In fact, all the eigenvalues of V are equal to zero, and, in particular, $V\Delta P_1=0$. Now Theorem 1.6.1 gives (6.6).

Corollary 1.6.2 *Let D and V be the diagonal and the nilpotent parts of $A\in B(\mathbf{C}^n)$, respectively. Then the equality*

$$R_\lambda(A)=R_\lambda(D)\overrightarrow{\prod_{2\leq k\leq n}}[I+(\lambda-\lambda_k(A))^{-1}V\Delta P_k] \tag{6.7}$$

is valid for all regular λ.

It is true, due to (3.1)

$$R_\lambda(A)=(A-\lambda I)^{-1}=(D+V-\lambda I)^{-1}=R_\lambda(D)(I+VR_\lambda(D))^{-1}.$$

But $VR_\lambda(D)$ is a nilpotent operator. Take into account that

$$R_\lambda(D)\Delta P_k=(\lambda_k(A)-\lambda)^{-1}\Delta P_k.$$

Now (6.6) ensures the relation (6.7).

1.7 A relation between the determinant and the resolvent of a matrix

1.7.1 Statement of the results

Let $\chi(z)$ be the function of a complex argument z defined by

$$\chi(z) = \max\{|z|, 1\}. \tag{7.1}$$

Theorem 1.7.1 *For any* $A \in B(\mathbf{C}^n)$ *and a regular point* λ *of* A *the following inequality is true*

$$\|R_\lambda(A) det(A - I\lambda)\| \leq \omega(I\lambda - A)G(A), \tag{7.2}$$

where

$$\omega(I\lambda - A) = \prod_{j=1}^{n} \chi(\lambda_j(A) - \lambda)$$

and

$$G(A) = G_n(A) = \sum_{k=0}^{n-1} g^k(A)\gamma_{n,k}(A).$$

The proofs are given in the next subsection.

Corollary 1.7.2 *For any* $A \in B(\mathbf{C}^n)$ *and a regular* λ *the relation*

$$|det(A - I\lambda)| \leq \|A - I\lambda\|\omega(A - \lambda I)G(A) \tag{7.3}$$

holds.

Corollary 1.7.3 *For any* $A \in B(\mathbf{C}^n)$ *and a regular* λ *the following inequality is valid*

$$\|R_\lambda(A)\| \leq \frac{G(A)}{\prod_{k=1}^{n} \psi(\lambda - \lambda_k(A))}, \tag{7.4}$$

where

$$\psi(z) = min\{1, |z|\} \text{ for a complex number } z.$$

Notice that if A is a normal operator, then $g(A) = 0$, (see Section 1.3) and therefore $G(A) = 1$.

Theorem 1.7.1 is exact. In fact, let A be a unitary operator. Since any unitary operator is normal, and $|\lambda_k(A)| = 1$, then $\chi(\lambda_k(A)) = 1$. We have due to (7.3) with $\lambda = 0$ the relations

$$|det A| = 1 \le \|A\|.$$

But the norm of a unitary operator equals 1. Thus we arrive at the equality

$$|det A| = \|A\| = 1.$$

One more example. Let A be a negative Hermitian matrix with $\min_k \lambda_k(A) = 0$. Then $g(A) = 0$ and Theorem 1.7.1 gives

$$\|(I - A)^{-1} det(I - A)\| \le \prod_{k=1}^n (1 - \lambda_k(A)) = det(I - A),$$

because $1 - \lambda_k(A) \ge 1$ in this case. But since A is a Hermitian matrix, we can write down

$$\|(I - A)^{-1}\| = \frac{1}{\min_k(1 - \lambda_k(A))} = 1.$$

Consequently, inequality (7.2) is attained in this case, as well.

Note that

$$\omega(I\lambda - A) \le \prod_{j=1}^n c_j$$

where

$$c_j = \max\{1, |\lambda_j(A)| + |\lambda|\}.$$

1.7.2 Proofs of Theorem 1.7.1 and its Corollaries

Let $\{e_k\}$ be the Schur basis of A, i.e. the orthonormed basis such that A is represented with respect to $\{e_k\}$ by an upper triangular matrix.

Lemma 1.7.4 *Let $I - A$ be an invertible operator in $\mathbf{C}^n$. Then the relation*

$$\|(I - A)^{-1} det(I - A)\| \le \omega(I - A)\|(I - |V|)^{-1}\| \qquad (7.5)$$

is true, where $|V|$ is the matrix whose elements with respect to the Schur basis are moduli of elements of the nilpotent part V of A with respect to this basis.

That is elements of $|V|$ are

$$(|V|e_k, e_j) = |(Ve_k, e_j)| = |(Ae_k, e_j)| \text{ for } j < k$$

and

$$(|V|e_k, e_j) = 0 \text{ for } j \geq k.$$

Proof: Due to Theorem 1.6.1 we can write down

$$(I - A)^{-1} = \overrightarrow{\prod_{1 \leq k \leq n}} (I + (1 - \lambda_k(A))^{-1}A\Delta P_k).$$

We thus get

$$(I - A)^{-1} = det^{-1}(I - A) \overrightarrow{\prod_{1 \leq k \leq n}} (I(1 - \lambda_k(A)) + A\Delta P_k).$$

Put

$$K \equiv (I - A)^{-1}det(I - A).$$

Due to the equality

$$A\Delta P_k = \lambda_k(A)\Delta P_k + V\Delta P_k$$

we have

$$K = \overrightarrow{\prod_{1 \leq k \leq n}} (I - \lambda_k(A)\overline{P}(k) + V\Delta P_k), \qquad (7.6)$$

where $\overline{P}(k) = I - \Delta P_k$.

Since

$$(I + V\Delta P_k)(I - \lambda_k(A)\overline{P}(k)) = I - \lambda_k(A)\overline{P}(k) + V\Delta P_k,$$

we conclude

$$K = \overrightarrow{\prod_{1 \leq k \leq n}} (I + V\Delta P_k)(I - \lambda_k(A)\overline{P}(k)).$$

Set

$$K_j = \prod_{1 \leq k \leq j}^{\rightarrow} (I + V \Delta P_k)(I - \lambda_k(A)\overline{P}(k)).$$

Then

$$K_n = K = K_{n-1}(I + V \Delta P_n)(I - \lambda_n(A)\overline{P}(n)).$$

It is obvious that

$$|(I - \lambda_n(A)\overline{P}(n))x| \leq |\overline{P}(n) - \lambda_n(A)\overline{P}(n)||x| + \Delta P_n|x|,$$

where the inequality is understood for each coordinate with respect to the Schur basis, and the symbol $| \, . \, |$ again means that we take moduli of elements of matrices and vectors with respect to this one.

Exploiting Lemma 1.3.9, we have

$$\|Kx\| \leq \| \, |K_{n-1}|(I + |V|\Delta P_n)(|1 - \lambda_n(A)|\overline{P}(n) + \Delta P_n)|x| \, \|.$$

Further, it is clear that

$$(|1 - \lambda_n(A)|\overline{P}(n) + \Delta P_n)|x| \leq \chi(1 - \lambda_n(A))|x|.$$

Thus,

$$\|Kx\| \leq \chi_n\| \, |K_{n-1}|(I + |V|\Delta P_n)|x| \, \|.$$

Here

$$\chi_k = \chi(1 - \lambda_k(A)).$$

Similarly, due to the equality

$$K_{n-1} = K_{n-2}(I + V \Delta P_{n-1})(I - \lambda_{n-1}(A)\overline{P}(n-1))$$

we obtain

$$\|Kx\| \leq \chi_{n-1}\chi_n\|K_{n-2}(I + |V|\Delta P_{n-1})(I + |V|\Delta P_n)|x|\|.$$

Repeating these arguments, we arrive at the inequality

$$\|K\| \leq \prod_{1 \leq j \leq n} \chi_j\| \prod_{1 \leq k \leq n}^{\rightarrow} (I + |V|\Delta P_k)\|. \tag{7.7}$$

According to Theorem 1.7.1

$$\overrightarrow{\prod_{1 \le k \le n}} (I + |V| \Delta P_k) = (I - |V|)^{-1}.$$

From this and (7.7) the result follows. $\square$

Corollary 1.7.5 *Let $I - A$ be an invertible matrix in C^n. Then the estimate*

$$\|det(I - A)(I - A)^{-1}\| \le \omega(I - A)G(A) \qquad (7.8)$$

holds.

In fact, Theorem 1.2.2 implies

$$\|(I - |V|)^{-1}\| \le \sum_{k=0}^{n} \gamma_{k,n} N^k(V).$$

Take into account that $N(V) = N(|V|)$. Due to Corollary 1.3.7 $N(V) = g(A)$. Thus we arrive at the inequality

$$\|(I - |V|)^{-1}\| \le \sum_{k=0}^{n} \gamma_{k,n} g^k(A) = G(A).$$

Now Lemma 1.7.4 gives the result. $\square$

Proof of Theorem 1.7.1: Replacing in (7.8) A by $A - (\lambda - 1)I$, and using Corollary 1.3.8 we arrive at the relation

$$\|det(\lambda I - A)R_\lambda(A)\| \le G(A) \prod_{j=1}^{n} \chi(\lambda - \lambda_j(A)).$$

As claimed. $\square$

Proof of Corollary 1.7.2: The assertion of Theorem 1.7.1 means that

$$\|(I\lambda - A)^{-1}h\| \|det(A - I\lambda)\| \le \omega(I\lambda - A)G(A)\|h\|$$

for any vector h. Substituting

$$x = (I\lambda - A)^{-1}h$$

into this relation we easily get the stated result.□

Proof of Corollary 1.7.3: We have the equality

$$\frac{\chi(z)}{|z|} = \frac{1}{\psi(z)}$$

which gives

$$\frac{\chi(\lambda - \lambda_k(A))}{|\lambda - \lambda_k(A)|} = \frac{1}{\psi(\lambda - \lambda_k(A))}.$$

From this and Theorem 1.7.1 the inequality (7.4) follows. □

1.8 Notes

Section 1.1 The quantity $g(A)$ was introduced by Henrici (1962) and independently by Gil'(1983a).

Section 1.2 I.M. Gel'fand and G.E. Shilov (1958) established the estimate

$$\|f(A)\| \leq \sum_{k=0}^{n-1} \sup_{\lambda \in co(A)} |f^{(k)}(\lambda)|(2\|A\|)^k.$$

Carleman (see (Dunford and Schwartz, 1963)) derived the inequality

$$\|R_\lambda(A) \prod_{k=1}^{n}(1 - \lambda^{-1}\lambda_k(A))exp[\lambda^{-1}\lambda_k(A)]\| \leq$$

$$|\lambda|exp[1 + N^2(A\lambda^{-1})/2].$$

About the estimation for the matrix exponent see also Coppel (1978). Theorems 1.2.1 and 1.2.2 are established in the paper by Gil'(1993a). Corollary 1.2.3 is derived in Gil' (1983a), (see also Gil'(1979a)) . Corollary 1.2.4 is established in Gil' (1983b).

Section 1.5 Theorem 1.6.1 is published in Gil'(1993a). It improves and supplements Schur's and Brown's inequalities (see Marcus and Minc (1964)). A very interesting inequality for eigenvalues of matrices was derived by Kress, de Vries and Wegman (1974).

Section 1.6 Theorem 1.6.1 is published in the more general situation in Gil' (1973b).

Section 1.7 About other relations between the determinant and the norm for the resolvent see the book by Gohberg and Krein (1969, Chapter 5).

Chapter 2

Functions of Compact Operators

This chapter is devoted to functions of compact operators.

In Section 2.1 and Section 2.2 we recall some basic notions of the theory of compact operators.

In Section 2.3 an estimation for the norm of the powers of a Volterra Hilbert-Schmidt operator is derived.

Section 2.4 is concerned with an estimation for the norm of the resolvent of a Hilbert-Schmidt operator.

Section 2.5 analyzes operators whose powers are Hilbert-Schmidt operators. Estimations for the norm of the resolvent are given.

In Section 2.6 we investigate compact operators belonging to a Neumann-Schatten ideal. Estimations for the norm of the resolvent are derived.

Section 2.7 is concerned with an estimation for the norm of a regular function of a Hilbert-Schmidt operator.

The multiplicative structure of the resolvent of a compact operator is treated in Section 2.8. Namely, we obtain a representation of the resolvent by a product of simple operators which extends the spectral representation of the resolvent of a normal compact operator.

By the multiplicative representation of the resolvent, in Section 2.9 a relation between the determinant and the norm of the resolvent is established for a Hilbert-Schmidt operator with a finite trace.

In Section 2.10 inequalities for nilpotent parts of operators belonging to the Neumann-Schatten ideals are derived.

Relations for eigenvalues of Hilbert-Schmidt operators are established in Section 2.11.

2.1   Terminology

In this section we recall very briefly some basic notions of the theory of operators in a Hilbert space. More details can be found in any textbook on Hilbert spaces (see for instance, Ahiezer and Glazman (1981), Dunford and Schwartz (1963) etc).

Let H be a Hilbert space with the scalar product $(.,.)$ and the norm

$$\|h\| = \sqrt{(h,h)}, \ h \in H.$$

A sequence $\{h_n\}$ of elements of H converges *strongly* (in the norm) to $h \in H$ if

$$\lim_{n \to \infty} \|h_n - h\| = 0.$$

A linear operator A acting in H is bounded if there is a constant a such that

$$\|Ah\| \le a\|h\| \text{ for all } h \in H.$$

The quantity

$$\|A\| = \sup_{h \in H} \frac{\|Ah\|}{\|h\|}$$

is called the norm of A.

A sequence $\{A_n\}$ of bounded linear operators *converges strongly* to an operator A if the sequence of elements $\{A_n h\}$ strongly converges to Ah for every $h \in H$.

A sequence $\{A_n\}$ of bounded linear operators converges *in the uniform operator topology* (in the operator norm) to an operator A if

$$\lim_{n \to \infty} \|A_n - A\| = 0.$$

A bounded linear operator A^* is called adjoint to A if

$$(Af, g) = (f, A^*g) \text{ for every } h, g \in H. \tag{1.1}$$

The relation

$$\|A\| = \|A^*\|$$

is true.

A bounded operator A *is a selfadjoint* one if $A = A^*$.

A is *a unitary operator* if

$$AA^* = A^*A = I.$$

Throughout this chapter $I \equiv I_H$ is the identity operator in H.

A selfadjoint operator A is positive (negative) if

$$(Ah, h) \geq 0 \ ((Ah, h) \leq 0) \text{ for every } h \in H.$$

A bounded linear operator satisfying the relation

$$AA^* = A^*A \tag{1.2}$$

is called *a normal operator*.

It is clear that unitary and selfadjoint operators are examples of normal ones. The operator $B \equiv A^{-1}$ is the inverse one to A if

$$AB = BA = I.$$

A point λ of the complex plane is said to be a regular point of an operator A if the operator

$$R_\lambda(A) \equiv (A - I\lambda)^{-1}$$

(the resolvent) exists and is bounded.

The complement of all regular points of A in the complex plane is the *spectrum* of A. The spectrum of A is denoted by $\sigma(A)$.

The quantity

$$d(A) = sup_{s \in \sigma(A)}|s|$$

is called the *spectral radius* of A. Gelfand's formula

$$d(A) = lim_{n \to \infty} \sqrt[n]{\|A^n\|}$$

is true.

An operator V is called a quasinilpotent one if its spectrum consists of zero, only. This means that

$$\lim_{n \to \infty} \sqrt[n]{\|V^n\|} = 0. \qquad (1.3)$$

If there is a nontrivial solution of the equation

$$Ae = \lambda(A)e,$$

where $\lambda(A)$ is a number, then this number is called an eigenvalue of the operator A, and $e \in H$ is an eigenvector corresponding to $\lambda(A)$.

Any eigenvalue is a point of the spectrum. An eigenvalue $\lambda(A)$ has the algebraic multiplicity $r \leq \infty$ if

$$dim(\cup_{k=1}^{\infty} ker(A - \lambda I)^k) = r.$$

The spectrum of a selfadjoint operator is real, the spectrum of a unitary operator lies on the unit circle.

An operator P is called *a projector* if $P^2 = P$. If, in addition, $P^* = P$, then it is called *an orthogonal projector* (an orthoprojector).

If in a Hilbert space there is a countable set whose closure coincides with whole space, then the space is said to be *separable* . Any separable Hilbert space H possesses an orthonormed basis. This means that there is a sequence $\{e_k \in H\}$ such that

$$(e_k, e_j) = 0 \text{ if } j \neq k \text{ and } (e_k, e_k) = 1$$

and any $h \in H$ can be represented as

$$h = \sum_{k=1}^{\infty} c_k e_k$$

with

$$c_k = (h, e_k), k = 1, 2, \ldots$$

Besides the series strongly converges.

A linear operator in a Hilbert space is a completely continuous (compact) one if it maps each bounded set into a compact one.

Any nonzero eigenvalue of a completely continuous operator has the finite multiplicity. The spectrum is either finite or sequence $\{\lambda_j(A)\}$ or the eigenvalues of A converges to zero, only.

Any completely continuous normal operator A can be represented in the form

$$A = \sum_{k=1}^{\infty} \lambda_k(A) E_k, \qquad (1.4)$$

where $\lambda_k(A)$ are eigenvalues of A repeated according to their multiplicities, and E_k are eigenprojectors of A, i.e. the orthoprojectors defined by

$$E_k = (., e_k) e_k,$$

where e_k are the eigenvectors. The last equality means

$$E_k h = (h, e_k) e_k \text{ for any } h \in H.$$

A completely continuous positive selfadjoint operator has non-negative eigenvalues, only.

Let A be a completely continuous positive selfadjoint operator represented by (1.4), then the operator

$$A^{1/2} \equiv \sum_{k=1}^{\infty} \sqrt{\lambda_k(A)} E_k$$

with the positive roots $\sqrt{\lambda_k(A)}$ is called *a square root of A*.

A completely continuous quasinilpotent operator is called a Volterra operator.

Let $\{e_k\}$ be an orthonormed basis in H, and let an operator A satisfy the following condition: the series

$$\sum_{k=1}^{\infty}(Ae_k, e_k)$$

converges, then the sum of this series is called *the Trace of the operator A* and is written

$$Trace A = Tr A = \sum_{k=1}^{\infty}(Ae_k, e_k).$$

If for linear operators A and B

$$Tr(AB) < \infty,$$

then as in the case of finite matrices

$$Tr(AB) = Tr(BA).$$

The following relationship (Lidskij's theorem)

$$Tr A = \sum_{k=1}^{\infty}\lambda_k(A)$$

is true. An operator A satisfying the condition

$$Tr(A^*A)^{1/2} < \infty$$

is called *a nuclear operator.*

An operator A satisfying the relation

$$Tr(A^*A) < \infty$$

is said to be *a Hilbert-Schmidt operator* .

Let A be an arbitrary completely continuous operator, then an eigenvalue $\lambda_j((A^*A)^{1/2})$, $j = 1, 2, \dots$ of the operator $(A^*A)^{1/2}$ are called *singular numbers* (s-numbers) of A and are denoted $s_j(A)$. That is,

$$s_k(A) \equiv \lambda_k((A^*A)^{1/2}).$$

Enumerate singular numbers of A taking into account their multiplicity and in decreasing order.

The set of completely continuous operators acting in a Hilbert space and satisfying for some $p \geq 1$ the condition

$$N_p(A) \equiv \left(\sum_{k=1}^{\infty} s_k^p(A) \right)^{1/p} < \infty,$$

is called *the Neumann-Schatten ideal and is denoted by* C_p, $N_p(.)$ *is called the norm of the ideal* C_p.

It is not hard to show that

$$N_p(A) = \sqrt[p]{Tr(AA^*)^{p/2}}.$$

Thus C_1 is the ideal of nuclear operators (*the Trace class*), and C_2 is the ideal of Hilbert-Schmidt operators.

$N_2(A)$ is called the *Hilbert-Schmidt norm*. Sometimes we will omit index 2 of the Hilbert-Schmidt norm, i.e.

$$N(A) \equiv N_2(A) = \sqrt{Tr(A^*A)}.$$

For any orthonormed basis $\{e_k\}$ we can write down

$$N_2(A) = N(A) = \left(\sum_{k=1}^{\infty} \|Ae_k\|^2 \right)^{1/2}.$$

This equality is equivalent to the following

$$N(A) = \left(\sum_{j,k=1}^{\infty} |a_{jk}|^2 \right)^{1/2}$$

where

$$a_{jk} = (Ae_k, e_j) \; ; j, k = 1, 2, \ldots,$$

are entries of a Hilbert-Schmidt operator A in an orthonormal basis $\{e_k\}$.

For $p, q \geq 1$ the following propositions are true (the proofs can be found in the books by Dunford and Schwartz (1963), Gohberg and Krein (1969), and Pietsch (1988)).

If $A \in C_p$ then also $A^* \in C_p$.

If $A \in C_p$ and B is a bounded linear operator, then both AB and BA belong to C_p. Moreover,

$$N_p(AB) \le N_p(A)\|B\| \text{ and } N_p(BA) \le N_p(A)\|B\|.$$

The following inequality:

$$\sum_{j=1}^{n} |\lambda_j(A)|^p \le \sum_{j=1}^{n} s_j^p(A) \; ; n = 1, 2, \ldots \tag{1.5}$$

is valid.

If $A \in C_p$, and $B \in C_q$, then $AB \in C_s$ with

$$\frac{1}{s} = \frac{1}{p} + \frac{1}{q}.$$

Moreover, the inequality

$$N_s(AB) \le N_p(A)N_q(B) \tag{1.6}$$

holds.

For more details about ideals of completely continuous operators see for example Dunford and Schwartz (1963), Gohberg and Krein (1969), Gohberg and Krein (1970) and Pietsch (1988).

2.2   Schur's basis

A subspace $H_1 \subset H$ is said to be *an invariant subspace* of a linear operator A if the relation $h \in H_1$ implies $Ah \in H_1$. The projector P of H onto H_1 satisfies the relation $PAP = AP$.

H_1 is *a nontrivial invariant subspace* of A if it is an invariant subspace of A and $H_1 \ne H$, and $H_1 \ne \{0\}$. This means that the projector P of H onto H_1 satisfies the additional relations $P \ne I$, and $P \ne 0$.

As is well known (see for instance Gohberg and Krein (1970), Beauzamy (1988)), any completely continuous operator A possesses nontrivial invariant subspaces.

Let $KerA$ denote the kernel of an operator A :

$$KerA = \{h \in H : Ah = 0\},$$

and let λ be an eigenvalue of A, then

$$Ker(A - \lambda I)$$

is said to be an *eigenspace of A corresponding to λ* .

A vector $h \in H$ is called *a root vector* of an operator A corresponding to λ if

$$(A - I\lambda)^m h = 0$$

for some natural m.

The set of all the root vectors corresponding to λ is said to be *the root subspace of A corresponding to λ*.

Let us suppose that the system of the root vectors of a compact linear operator A acting in H is complete in H. This means that the closed linear span of all the root spaces of the operator A coincides with H. Then as well known (see Gohberg and Krein (1969), Section 1.4) there is an orthonormed basis $\{e_k\}$ by which A can be represented by a triangular matrix:

$$Ae_k = \sum_{j=1}^{k} a_{jk}e_j \text{ with } a_{jk} = (Ae_j, e_k),$$

This basis is called *Schur's basis* of the operator A.

Theorem 2.2.1 *Any completely continuous operator A can be decomposed as*

$$A = S + V \tag{2.1}$$

with a normal operator S and a Volterra (completely continuous quasinilpotent) operator V. Besides,

$$\sigma(S) = \sigma(A) \tag{2.2}$$

and all invariant subspaces of both V and S are invariant subspaces of A.

To prove this theorem we recall the following well-known result (see for instance (Gohberg and Krein, 1969), Lemma 1.4.2).

Lemma 2.2.2 *Let $M \neq H$ be the closed linear span of all the root vectors of a linear compact operator A, and let Q_A be the orthogonal projector of H onto $M^\perp$, where $M^\perp$ is the orthogonal complement of M in H. Then $Q_A A Q_A$ is a Volterra operator.*

Corollary 2.2.3 *Any eigenvector f of a compact linear operator A corresponding to a non-zero eigenvalue $\lambda(A)$, belongs to the closed linear span M of all the root vectors of A.*

Indeed, since M is an invariant subspace and due to Lemma 2.2.2 A can be represented by the matrix

$$A = \begin{pmatrix} A_1 & A_{12} \\ 0 & V_1 \end{pmatrix} \tag{2.3}$$

where $V_1 = Q_A A Q_A$ is a Volterra operator in $Q_A H$ and a completely continuous operator A_1 acting in $M = (I - Q_A)H$ has a set of the root vectors complete in M. Besides, $A_{12} = (I - Q_A)A Q_A$.

On the contrary suppose $f = f_1 + f_2$, where $f_1 \in M$ and $f_2 \in M^\perp \neq 0$. Then we must write

$$(Af)_1 = A_1 f_1 + A_{12} f_2 = \lambda(A) f_1 \text{ and } (Af)_2 = V_1 f_2 = \lambda(A) f_2,$$

where $\lambda(A)$ is the corresponding eigenvalue and $(Af)_j$ denotes a coordinate of Af. This is impossible, because V_1 is a Volterra operator. Thus $f_2 = 0$. As claimed.

Lemma 2.2.4 *Let Q_A be the same as at Lemma 2.2.2. Then the operator $V_0 = A Q_A$ is a Volterra one.*

Proof: We have $V_0 = V_1 + A_{12}$, where as above,

$$V_1 = Q_A A Q_A \text{ and } A_{12} = (I - Q_A)A Q_A.$$

Evidently, $A_{12}^2 = 0$ and $V_1 A_{12} = 0$. Hence,

$$(V_1 + A_{12})^n = V_1^n + A_{12} V_1^{n-1}.$$

This implies the result, since V_1 is a Volterra operator by Lemma 2.2.2. □

Proof of Theorem 2.2.1: Let A at the beginning have a set of root vectors complete in H. Denote by P_k, $k = 0, 1, 2, ...$ the orthoprojectors onto the invariant subspaces of A:

$$P_k = \sum_{j=1}^{k} (., .e_j)e_j, \ k = 1, 2, ... \tag{2.4}$$

where e_k is the corresponding Schur basis. Thus,

$$P_k A P_k = A P_k, \ k = 1, 2, \dots, P_\infty = I$$

and

$$0 \equiv Range(P_0) \subset Range(P_1) \subset Range(P_2) \subset \dots .$$

In this case

$$S e_k = a_{kk} e_k \text{ and } V e_k = \sum_{j=1}^{k-1} a_{jk} e_j. \tag{2.5}$$

Eigenvalues of S and A coincide (Gohberg and Krein, 1969, Sec. 1.4):

$$\lambda_k(A) = \lambda_k(S) = a_{kk}, \ k = 1, 2, \dots .$$

We have

$$S = \sum_{k=1}^{\infty} \lambda_k(A) \Delta P_k.$$

Here and below

$$\Delta P_k = P_k - P_{k-1}, \ k = 1, 2, \dots .$$

Besides,

$$P_k V P_k = V P_k, \ k = 1, 2, \dots .$$

Let now the set M of all root vectors of the operator A not coinciding with H.

We use (2.3) where A_1 has a set of the root vectors complete in M. As proved above, there are a normal operator S_1 and a Volterra operator V_2 such that the equation $A_1 = S_1 + V_2$ holds.

Put according to (2.3) $A = S + V$ with

$$S = \left(\begin{array}{cc} S_1 & 0 \\ 0 & 0 \end{array} \right)$$

and

$$V = \left(\begin{array}{cc} V_2 & A_{12} \\ 0 & V_1 \end{array} \right).$$

It is clear that S is a normal operator, and V is a Volterra one. We leave it to the reader to verify that $\sigma(S) = \sigma(S_1) \cup \{0\} = \sigma(A)$ and all the invariant subspaces of V and S are invariant subspaces of A.□

We will call equality (2.1) the triangular representation of a completely continuous operator A. Besides, the operators S and V will be called the diagonal part and the nilpotent part of A, respectively.

2.3 Powers of a Hilbert-Schmidt Volterra operator

Lemma 2.3.1 *Let V be a Hilbert-Schmidt completely continuous quasinilpotent (Volterra) operator acting in a separable Hilbert space H. Then the inequality*

$$\|V^k\| \leq \frac{N^k(V)}{\sqrt{k!}} \text{ for any natural } k \qquad (3.1)$$

is true.

Proof: Let $L^2[0,1]$ be the Hilbert space of scalar-valued functions defined on $[0,1]$ with the scalar product

$$(h,g) = \int_0^1 h(s)\overline{g}(s)ds.$$

Consider in $L^2[0,1]$ a Volterra Hilbert-Schmidt integral operator W defined by

$$(Wh)(t) = \int_0^t K(t,s)h(s)ds, \ h \in L^2[0,1],$$

where $K(t,s)$ is a scalar kernel. It is well known that the kernel satisfies the condition

$$\int_0^1 \int_0^t |K(t,s)|^2 ds dt = N^2(W) < \infty.$$

cf. Gohberg and Krein (1970). Here as above we set $N(.) = N_2(.)$.

Employing Schwarz's inequality, we have

$$\|Wh\|^2 = \int_0^1 |\int_0^t K(t,s)h(s)ds|^2 dt \leq$$

$$\int_0^1 \int_0^t |K(t,s)|^2 ds \int_0^t |h(s_1)|^2 ds_1 dt .$$

Setting

$$w(t) = \int_0^t |K(t,s)|^2 ds, \tag{3.2}$$

one can rewrite the latter inequality in the shape

$$\|Wh\|^2 \leq \int_0^1 w(t) \int_0^t |h(s_1)|^2 ds_1 dt.$$

Using this inequality, we obtain

$$\|W^k h\|^2 \leq \int_0^1 w(t) \int_0^t |W^{k-1} h(s_1)|^2 ds_1 dt ,$$

Once more apply Schwarz's inequality :

$$\|W^k h\|^2 \leq \int_0^1 w(t) \int_0^t w(s_1) \int_0^{s_1} |W^{k-2} h(s_2)|^2 ds_2 ds_1 dt.$$

Repeating these arguments, we arrive at the inequality

$$\|W^k h\|^2 \leq \int_0^1 w(t) \int_0^t w(s_1) \int_0^{s_1} w(s_2) \cdots$$

$$\int_0^{s_{k-2}} w(s_{k-1}) \int_0^{s_{k-1}} |h(s_k)|^2 ds_k \ldots ds_2 ds_1 dt.$$

Taking

$$\|h\|^2 = \int_0^1 |h(s)|^2 ds = 1,$$

we get

$$\|W^k\|^2 \le \int_0^1 w(t) \int_0^t w(s_1) \ldots \int_0^{s_{k-2}} w(s_{k-1}) ds_{k-1} \ldots ds_2 ds_1 dt. \quad (3.3)$$

It is simple to see that

$$\int_0^1 w(s_1) \ldots \int_0^{s_{k-1}} w(s_k) ds_k \ldots ds_1 =$$

$$\int_0^1 \int_0^{z_1} \ldots \int_0^{z_{k-1}} dz_k dz_{k-1} \ldots dz_1 = \frac{m^k}{k!}.$$

Here

$$z_k = z_k(s_k) \equiv \int_0^{s_k} w(s_{k-1}) ds_{k-1},$$

and

$$m = \int_0^1 w(s) ds.$$

Thus (3.3) gives

$$\|W^k\|^2 \le \frac{(\int_0^1 w(s) ds)^k}{k!}.$$

But according to (3.2)

$$m = \int_0^1 w(s) ds = N^2(W).$$

Thus,

$$\|W^k\|^2 \le \frac{N^{2k}(W)}{k!} \quad (3.4)$$

and for W we have proved the inequality (3.1).

Due to the well-known Theorem 2.2 from the book by Gohberg and Krein (1970), there exists a nonessential extension of the operator V which is unitarily equivalent to a some Volterra Hilbert-Schmidt operator V_0 acting in the space $L^2([0,1], \mathbf{C}^n)$ of vector-valued functions defined on [0,1] with the scalar product

$$(h, g) = \int_0^1 \sum_{k=1}^n h_k(s) \overline{g_k}(s) ds$$

where $h_k(s), g_k(s)$ are the coordinates of the functions $h, g \in L^2([0,1], \mathbf{C}^n)$, respectively.

Applying our arguments above to the operator V_0, we obtain the following inequality which is similar to (3.4):

$$\|V_0^k\|^2 \leq \frac{N^{2k}(V_0)}{k!}.$$

Since V and V_0 are unitarily equivalent,

$$\|V^k\| = \|V_0^k\|, \text{ and } N(V) = N(V_0).$$

Thus, we get the inequality (3.1). □

2.4   Resolvents of Hilbert-Schmidt operators

Let A be a Hilbert-Schmidt operator, i.e $A \in C_2$. The following quantity plays a key role in this section

$$g(A) = [N^2(A) - \sum_{k=1}^{\infty} |\lambda_k(A)|^2]^{1/2}, \qquad (4.1)$$

where $N(A) = N_2(A)$ is the Hilbert-Schmidt norm of A, again.

Since

$$\sum_{k=1}^{\infty} |\lambda_k(A)|^2 \geq |\sum_{k=1}^{\infty} \lambda_k^2(A)| = |TrA^2|,$$

one can write down

$$g^2(A) \leq N^2(A) - |TrA^2|.$$

If A is a normal Hilbert-Schmidt operator, then as in the finite dimension case $g(A) = 0$ (see Lemma 2.4.6 below).

We will also prove the inequality

$$g^2(A) \leq \frac{1}{2} N^2(A_I). \qquad (4.2)$$

(see Corollary 3.4.6 below). Here $A_I = (A - A^*)/2i$.

Theorem 2.4.1 *Let A be a linear Hilbert-Schmidt operator. Then the estimation*

$$\|R_\lambda(A)\| \leq \sum_{k=0}^{\infty} \frac{g^k(A)}{\rho^{k+1}(A, \lambda)\sqrt{k!}} \text{ for all regular } \lambda \qquad (4.3)$$

is valid.

Here $\rho(A, \lambda)$ is the distance between the spectrum $\sigma(A)$ of A and a complex point λ, again.

First, we will prove a few preliminary results. Let M be a set in the complex plane and let $\epsilon > 0$. By $S(M, \epsilon)$ we denote the ϵ-neighborhood of M. That is,

$$dist\{M, S(M, \epsilon)\} \leq \epsilon.$$

Lemma 2.4.2 *Let A be a bounded operator, and let $\epsilon > 0$. Then there is a $\delta > 0$ such that if a bounded operator B satisfies the condition*

$$\|A - B\| \leq \delta,$$

then $\sigma(B)$ lies in $S(\sigma(A), \epsilon)$ and

$$\|R_\lambda(A) - R_\lambda(B)\| \leq \epsilon$$

for any λ which does not belong to $S(\sigma(A), \epsilon)$.

For the proof of this lemma we refer the reader to Dunford and Schwartz (1958, p. 584).

Lemma 2.4.3 *Let $V \in C_p, p \geq 1$ be a Volterra operator. Then there exists a sequence $\{W_n\}$ of compact nilpotent operators having finite dimensional ranges such that*

$$N_p(W_n - V) \to 0 \ as \ n \to \infty.$$

Proof: It is well known (see (Dunford and Schwartz, 1963), Lemma XI.9.10) that there exists a sequence $\{B_n\}$ of compact operators having finite dimension ranges such that

$$N_p(B_n - V) \to 0 \ as \ n \to \infty.$$

Since B_n has a finite dimensional range, it has the triangular representation

$$B_n = D_n + W_n$$

with the diagonal part D_n and the nilpotent part W_n. Moreover, $\sigma(D_n) = \sigma(B_n)$. According to Lemma 2.4.2, for a given $\varepsilon > 0$ and a sufficiently large n we have

$$\sigma(D_n) = \sigma(B_n) \subset S(\sigma(V), \varepsilon).$$

But $\sigma(V) = \{0\}$. Thus

$$\|D_n\| = max_k |\lambda_k(D_n)| \leq \varepsilon,$$

since D_n is normal. We can assert that

$$\|V - W_n\| \to 0 \text{ as } n \to \infty.$$

But due to the well-known Theorem 3.6.3 (Gohberg and Krein, 1969) if a sequence $\{A_n\}$ of operators in H converges strongly to A and A belongs to C_p, then this sequence converges to A in the norm $N_p(.)$. This proves the result since W_n are nilpotent. $\square$

Lemma 2.4.4 *Let $A \in C_p$, $p \geq 1$. Then there exists a sequence $\{A_n\}$ of compact operators such that A_n has an n-dimension range. Besides,*

$$\sigma(A_n) \subseteq \sigma(A). \tag{4.4}$$

Moreover $N_p(A_n - A) \to 0$ and

$$\sum_{k=1}^{n} |\lambda(A_n)|^p \to \sum_{k=1}^{\infty} |\lambda(A)|^p \text{ as } n \to \infty. \tag{4.5}$$

Proof: Due to Lemma 2.2.2 A can be represented in the form (2.3), where $V_1 = Q_A A Q_A$ is a Volterra operator in $Q_A H$, and a completely continuous operator A_1 acting in $(I - Q_A)H$ has a set of the root vectors complete in $(I - Q_A)H$. Besides, $A_{12} = Q_A A(I - Q_A)$.

Due to Corollary 2.2.3

$$\sigma(A) = \sigma(A_1) \cup \{0\}.$$

Let $A_1 = S_1 + V_{10}$ be the triangular representation of A_1 with the diagonal part S_1 and a nilpotent part V_{10}. As $\sigma(A_1) = \sigma(S_1)$, then

$$\sigma(A) = \sigma(S_1) \cup \{0\}.$$

Further, let P_n, $n = 1, 2, \ldots$ be orthogonal projectors onto invariant subspaces of A_1. Then $\sigma(A_1 P_n) = \sigma(S_1 P_n) \subseteq \sigma(A_1)$. It is clear that

$$N_p(A_1 P_n - A_1) \to 0 \text{ as } n \to \infty.$$

On the other hand, due to Lemma 2.2.4 $V_0 = AQ_A$ is a Volterra operator, and Lemma 2.4.3 asserts that there exists a sequence W_n of compact nilpotent operators having finite dimensional ranges such that $N_p(W_n - V_0)$ tends to zero as n tends to infinity.

Taking $A_n = A_1 P_n + W_n$, we arrive at the stated result. $\square$

Corollary 2.4.5 *Let A be a Hilbert-Schmidt operator. Then there exists a sequence $\{A_n\}$ of compact operators such that A_n has an n-dimension range, and the relations (4.4),*

$$N_2(A_n - A) \to 0, \text{ and } g(A_n) \to g(A) \text{ as } n \to \infty \qquad (4.6)$$

are true.

Proof of Theorem 2.4.1: Thanks to Corollary 2.4.5 there is a sequence $\{A_n\}$ of compact operators having finite dimensional ranges such that (4.4) and (4.6) are valid.

Due to Corollary 1.2.4

$$\|R_\lambda(A_n)\| \le \sum_{k=0}^{n-1} \frac{g^k(A_n)}{\rho^{k+1}(A_n, \lambda)\sqrt{k!}} \text{ for all regular } \lambda.$$

According to (4.4) $\rho(A_n, \lambda) \ge \rho(A, \lambda)$. Letting $n \to \infty$ in the latter relation, we arrive at the stated result. $\square$

Theorem 2.4.1 is precise. The inequality (4.3) becomes the equality

$$\|R_\lambda(A)\| = \rho^{-1}(A, \lambda)$$

if A is a normal operator since $g(A) = 0$ in this case.

Lemma 2.4.6 *Let V be the nilpotent part of a Hilbert-Schmidt operator A. Then the relation*

$$N(V) = g(A)$$

holds.

Proof: Let M be the closed linear span of the root vectors of A and $\{e_k\}$ be the Schur basis. Then we have

$$VS^* e_k = \sum_{j=1}^{k-1} a_{jk} \overline{a_{kk}},$$

where a_{jk} are the entries of A in Schur's basis, again (see Section 2.2). The diagonal of the operator VS^* in the basis $\{e_k\}$ consists of zeros, only. Thus, we can assert that VS^* is a Volterra operator.

Similarly, one can check that $S^* V$ is a Volterra operator. Therefore,

$$Tr(S^* V) = Tr(V^* S) = 0.$$

From this and from the triangular representation (2.1) the relation

$$N^2(A) \equiv Tr(A^* A) = Tr(S + V)^*(S + V) =$$

$$Tr(S^* S) + Tr(V^* V).$$

follows. But

$$Tr(S^* S) = \sum_{k=1}^{\infty} |\lambda_k(A)|^2.$$

This proves the result.$\square$

Corollary 2.4.7 *Let A be a Hilbert-Schmidt operator, and V be its nilpotent part. Then the inequality*

$$\|R_\lambda(A)\| \leq \sum_{k=0}^{\infty} \frac{N^k(V)}{\rho^{k+1}(A, \lambda)\sqrt{k!}} \quad \text{for all regular } \lambda,$$

is true.

Indeed, this result follows from Theorem 2.4.1 and Lemma 2.4.6.

2.5   Resolvents of operators with compact powers

Assume that for some positive integer $p \geq 1$ the following condition

$$A^p \text{ is a Hilbert-Schmidt operator} \tag{5.1}$$

holds. Note that A can, in general, be a noncompact operator. Below in this section we will give a relevant example.

Theorem 2.5.1 *Let (5.1) hold for some integer positive p. Then the following estimate is true:*

$$\|R_\lambda(A)\| \leq \|T_{\lambda,p}\| \sum_{k=0}^{\infty} \frac{g^k(A^p)}{\rho^{k+1}(A^p, \lambda^p)\sqrt{k!}} \text{ for all regular } \lambda, \qquad (5.2)$$

where

$$T_{\lambda,p} = \sum_{k=0}^{p-1} A^k \lambda^{p-k-1}, \qquad (5.3)$$

and

$$\rho(A^p, \lambda^p) = \inf_{t \in \sigma(A)} |t^p - \lambda^p|$$

is the distance between $\sigma(A^p)$ and the point λ^p.

Proof: We use the identity

$$A^p - I\lambda^p = (A - I\lambda)\sum_{k=0}^{p-1} A^k \lambda^{p-k-1} = (A - I\lambda)T_{\lambda,p}.$$

This implies

$$(A - I\lambda)^{-1} = T_{\lambda,p}(A^p - I\lambda^p)^{-1}. \qquad (5.4)$$

Thus,

$$\|(A - I\lambda)^{-1}\| \leq \|T_{\lambda,p}\| \, \|(A^p - I\lambda^p)^{-1}\|.$$

Applying Theorem 2.4.1 to the resolvent

$$(A^p - I\lambda^p)^{-1} = R_{\lambda^p}(A^p),$$

we obtain:

$$\|R_{\lambda^p}(A^p)\| \leq \sum_{k=0}^{\infty} \frac{g^k(A^p)}{\rho^{k+1}(A^p, \lambda^p)\sqrt{k!}} \text{ for all regular } \lambda$$

This completes the proof.□

Example 2.5.2 *Consider a noncompact operator satisfying the condition (5.1).*

Let H be an orthogonal sum of Hilbert spaces H_1 and H_2: $H = H_1 \oplus H_2$, and let A be a linear operator defined in H by the formula

$$A = \begin{pmatrix} B_1 & C \\ 0 & B_2 \end{pmatrix},$$

where B_1 and B_2 are bounded linear operators acting in H_1 and H_2, respectively, and a bounded linear operator C maps H_2 into H_1. Evidently A^2 is defined by the matrix

$$A^2 = \begin{pmatrix} B_1^2 & B_1 C + C B_2 \\ 0 & B_2^2 \end{pmatrix}.$$

If B_1, B_2 are compact operators and C is a noncompact one, then it is obvious that A^2 is a compact operator while A is a noncompact one.

2.6 Resolvents of Neumann-Schatten operators

Theorem 2.6.1 *Let a linear operator A acting in a separable Hilbert space satisfy the condition*

$$A \in C_{2r}, \text{ for some integer } r. \tag{6.1}$$

Then the inequality

$$\|R_\lambda(A)\| \leq \sum_{m=0}^{r-1} \sum_{k=0}^{\infty} \frac{(2N_{2r}(A))^{kr+m}}{\rho^{rk+m+1}(A,\lambda)\sqrt{k!}} \tag{6.2}$$

is true for all regular λ .

First, we will prove a few lemmata.

Let T be a compact operator with a set of root vectors complete in H. And let $\{e_k\}$ be a corresponding (orthogonal) Schur's basis. That is $(Te_k, e_j) = 0$ if $j \geq k$.

Note that if two operators have a common Schur's basis, then their sum and product have the same Schur basis as well. In the other words, the set of all bounded operators having a common Schur's basis is a ring.

Let now V_1 be a Volterra operator satisfying

$$(V_1 e_k, e_j) = 0 \text{ for } j \geq k. \tag{6.3}$$

That is, $\{e_k\}$ is the Schur basis for V_1 as well. Then it is easy to show that

$$(TV_1 e_k, e_j) = (V_1 T e_k, e_j) = 0 \text{ for } j \geq k.$$

Thus, TV_1 and $V_1 T$ are Volterra operators having a common Schur basis $\{e_k\}$. In other words the operators P_k defined by

$$P_j h = \sum_{k=1}^{j} (h, e_k) e_k \; ; h \in H, j = 1, 2, \dots$$

are orthogonal projectors onto the discrete chain of invariant subspaces of V_1.

Since the eigenvectors of a normal operator are the elements of its Schur basis, from our reasoning above follows:

Lemma 2.6.2 *Let a compact normal operator S_1 have eigenvectors $e_1, e_2, \dots$. And let them generate Schur's basis for a Volterra operator V_1, i.e. let equality (6.3) hold. Then $S_1 V_1$ and $V_1 S_1$ are Volterra operators with the same Schur's basis.*

Corollary 2.6.3 *Let V and S be the nilpotent and the diagonal parts of a completely continuous operator A, respectively, and let λ be a regular point of A. Then the operators $R_\lambda(S)V$ and $VR_\lambda(S)$ are Volterra ones.*

Lemma 2.6.4 *Let A be a compact operator and let V be its nilpotent part. If*

$$V \in C_{2r}, \text{ for some integer } r, \tag{6.4}$$

then the inequality

$$\|R_\lambda(A)\| \leq \sum_{m=0}^{r-1} \sum_{k=0}^{\infty} \frac{N_{2r}^{kr+m}(V)}{\rho^{rk+m+1}(A, \lambda)\sqrt{k!}} \text{ for all regular } \lambda \tag{6.5}$$

is true.

Proof: 1. Denote

$$B_\lambda = -V R_\lambda(S).$$

By the identity

$$(I - B_\lambda)(I + B_\lambda + \ldots + B_\lambda^{r-1}) = I - B_\lambda^r$$

we have

$$(I - B_\lambda)^{-1} = (I + B_\lambda + \ldots + B_\lambda^{r-1})(I - B_\lambda^r)^{-1}.$$

Due to Lemma 2.6.2 B_λ is a Volterra operator. Consequently, the series

$$(I - B_\lambda^r)^{-1} = \sum_{k=0}^{\infty} B_\lambda^{kr}$$

converges for all regular points of S. Thus,

$$(I - B_\lambda)^{-1} = (I + B_\lambda + \ldots + B_\lambda^{r-1})(I - B_\lambda^r)^{-1} = \sum_{m=0}^{r-1} \sum_{k=0}^{\infty} B_\lambda^{kr+m}. \quad (6.6)$$

According to (2.1) we can write down

$$R_\lambda(A) = R_\lambda(S)(I - B_\lambda)^{-1}.$$

Therefore, (6.6) implies

$$R_\lambda(A) = R_\lambda(S) \sum_{m=0}^{r-1} \sum_{k=0}^{\infty} B_\lambda^{kr+m}. \quad (6.7)$$

2. Taking into account that $\sigma(A) = \sigma(S)$, we can assert that $R_\lambda(S)$ is a bounded operator for all regular points λ of A. Hence $B_\lambda \in C_{2r}$. This implies that B_λ^r is a Hilbert-Schmidt operator, because

$$N_2(B_\lambda^r) \equiv N(B_\lambda^r) \le N_{2r}^r(B_\lambda) \quad (6.8)$$

(see Section 2.1). Now we use the estimate for powers of a Volterra Hilbert-Schmidt operator obtained in Section 2.3. It gives

$$\|B_\lambda^{rk}\| \le \frac{N^k(B_\lambda^r)}{\sqrt{k!}}.$$

Now (6.8) ensures the estimate

$$\|B_\lambda^{rk}\| \le \frac{N_{2r}^{kr}(B_\lambda)}{\sqrt{k!}}.$$

It is clear that

$$N_{2r}(B_\lambda) = N_{2r}(VR_\lambda(S)) \le N_{2r}(V)\|R_\lambda(S)\|.$$

Since S is normal and $\sigma(A) = \sigma(S)$,

$$\|R_\lambda(S)\| = \rho^{-1}(S,\lambda) = \rho^{-1}(A,\lambda). \tag{6.9}$$

Hence,

$$N_{2r}(B_\lambda) \le \frac{N_{2r}(V)}{\rho(A,\lambda)}. \tag{6.10}$$

Thus,

$$\|B_\lambda^{rk}\| \le \frac{N_{2r}^{kr}(V)}{\rho^{kr}(A,\lambda)\sqrt{k!}}.$$

Evidently,

$$\|B_\lambda^m\| \le N_{2r}^m(B_\lambda).$$

The relation (6.10) implies

$$\|B_\lambda^m\| \le \frac{N_{2r}^m(V)}{\rho^m(A,\lambda)}.$$

Consequently,

$$\|B_\lambda^{rk+m}\| \le \frac{N_{2r}^{kr+m}(V)}{\rho^{kr+m}(A,\lambda)\sqrt{k!}}.$$

From this, taking into account (6.7), we can write down

$$\|R_\lambda(A)\| \le \|R_\lambda(S)\| \sum_{m=0}^{r-1} \sum_{k=0}^{\infty} \frac{N_{2r}^{kr+m}(V)}{\rho^{kr+m}(A,\lambda)\sqrt{k!}}.$$

Now (6.9) yields the result.□

Lemma 2.6.5 *Let a linear operator A acting in a separable Hilbert space satisfy (6.1). Then inequality (6.5) is true, where V is the nilpotent part of A.*

Proof: According to (1.5) we have: $S \in C_{2r}$. Moreover,

$$N_p(S) \le N_p(A). \qquad (6.11)$$

Thus, the representation (2.1) implies the relation $V \in C_{2r}$. Now, the result follows from Lemma 2.6.4. $\square$

Lemma 2.6.6 *Let (6.1) be satisfied, then the nilpotent part of A satisfies the relations*

$$N_{2r}(V) \le N_{2r}(A) + N_{2r}(S) \le 2N_{2r}(A).$$

Proof: The triangular representation (2.1) gives

$$N_p(V) \le N_p(A) + N_p(S).$$

This and (6.11) prove the result.$\square$

The assertion of Theorem 2.6.1 immediately follows from Lemma 2.6.6 and Lemma 2.6.5.

2.7 Functions of a Hilbert-Schmidt operator

Let A be a bounded linear operator acting in a separable Hilbert space H, and f be a scalar-valued function which is analytic on a neighborhood of $\sigma(A)$. Let contour C consist of a finite number of rectifiable Jordan curves, oriented in the positive sense customary in the theory of complex variables. Suppose that C is the boundary of an open set $M \supset \sigma(A)$ and $M \cup C$ is contained in the domain of analycity of f. We define $f(A)$ by the equality

$$f(A) = -\frac{1}{2\pi i} \int_C f(\lambda) R_\lambda(A) d\lambda \qquad (7.1)$$

(see Dunford and Schwartz (1958, p.568)).

Theorem 2.7.1 *Let A be a Hilbert-Schmidt operator and let f be a holomorphic function in a neighborhood of the closed convex hull co(A) of the spectrum of A. Then the inequality*

$$\|f(A)\| \le \sum_{k=0}^{\infty} \sup_{\lambda \in co(A)} |f^{(k)}(\lambda)| \frac{g^k(A)}{(k!)^{3/2}} \qquad (7.2)$$

is valid.

Proof: Thanks to Corollary 2.4.5 there is a sequence $\{A_n\}$ of compact operators having finite dimensional ranges such that (4.4) and (4.6) hold. Corollary 1.2.3 implies

$$\|f(A_n)\| \leq \sum_{k=0}^{n-1} \sup_{\lambda \in co(A_n)} |f^{(k)}(\lambda)| \frac{g^k(A_n)}{(k!)^{3/2}}. \tag{7.3}$$

Due to the well-known Lemma VII.6.5 from Dunford and Schwartz (1958) we have

$$\|f(A_n) - f(A)\| \to 0 \text{ as } n \to \infty.$$

Letting $n \to \infty$ in (7.3), we arrive at the stated result. □

Theorem 2.7.1 is precise: the inequality (7.2) becomes the equality

$$\|f(A)\| = \sup_{\mu \in \sigma(A)} |f(\mu)|$$

if A is a normal operator and

$$\sup_{\lambda \in co(A)} |f(\lambda)| = \sup_{\lambda \in \sigma(A)} |f(\lambda)|,$$

because $g(A) = 0$ in this case.

Example 2.7.2 *Let A be a Hilbert-Schmidt operator, then the inequality*

$$\|A^m\| \leq \sum_{k=0}^{m} \frac{m! d^{m-k}(A) g^k(A)}{(m-k)!(k!)^{3/2}}$$

holds for every integer m. Recall that $d(A)$ is the spectral radius of the operator A.

Example 2.7.3 *Let A be a Hilbert-Schmidt operator, then the inequality*

$$\|e^{At}\| \leq e^{\alpha(A)t} \sum_{k=0}^{\infty} \frac{t^k g^k(A)}{(k!)^{3/2}} \text{ for all } t \geq 0$$

holds. Here $\alpha(A) = \sup \Re\sigma(A)$, again.

Lemma 2.7.4 *Let A be a Hilbert-Schmidt operator. Let also f be a function holomorphic in a neighborhood of co(A), then*

$$\|f(A)\| \leq \sum_{k=0}^{\infty} \sup_{\lambda \in co(A)} |f^{(k)}(\lambda)| \frac{N^k(V)}{(k!)^{3/2}},$$

where V is the nilpotent part of the operator A.

Proof: The result immediately follows from Theorem 2.7.1 and Lemma 2.4.6. □

2.8 Multiplicative structure of resolvents of compact operators

2.8.1 Operators with a complete system of root vectors

Let $X_1, X_2, ...$ be a sequence of bounded linear operators in H. For a natural n put

$$\prod_{1 \leq k \leq n}^{\rightarrow} (I + X_k) = (I + X_1)(I + X_2)...(I + X_n).$$

If there exists a limit in the operator norm of the products

$$\prod_{1 \leq k \leq n}^{\rightarrow} (I + X_k)$$

as n tends to ∞, then this limit we denote by

$$\prod_{1 \leq k \leq \infty}^{\rightarrow} (I + X_k).$$

Let now A be a complete continuous operator in H whose system of all the root vectors is complete in H and P_k, $k = 1, 2, ...$ be the sequence of orthogonal projectors onto the invariant subspaces defined by (2.4). Put

$$\Pi_n(\lambda) = \prod_{1 \leq k \leq n}^{\rightarrow} (I + \frac{A\Delta P_k}{\lambda - \lambda_k(A)}) \equiv (I + (\lambda_1(A) - \lambda)^{-1} A\Delta P_1)(I +$$
$$(\lambda_2(A) - \lambda)^{-1} A\Delta P_2)...(I + (\lambda_n(A) - \lambda)^{-1} A\Delta P_n).$$

for $\lambda \neq \lambda_k(A)$. Here $\Delta P_k = P_k - P_{k-1}, k = 1, 2, ... ; P_0 = 0$. That is,

$$\overrightarrow{\prod_{1 \leq k \leq \infty}} (I + \frac{A\Delta P_k}{\lambda - \lambda_k(A)})$$

is a limit in the operator norm of the sequence $\{\Pi_n(\lambda)\}$ as n tends to ∞.

Lemma 2.8.1 *Suppose that the system of all the root vectors of a compact linear operator A is complete in H. Then the equality*

$$\lambda R_\lambda(A) = - \overrightarrow{\prod_{1 \leq k \leq \infty}} (I + (\lambda - \lambda_k(A))^{-1} A\Delta P_k) \tag{8.1}$$

is valid for any regular λ.

Proof: Theorem 1.6.1 implies that for any compact operator A with the finite range n, the equality

$$\lambda R_\lambda(A) = -\Pi_n(\lambda) \tag{8.2}$$

is true. Now let A be an infinite dimensional operator in H. We denote $A_n = AP_n$. Then (8.2) gives for any $n < \infty$ the equality

$$\lambda(\lambda A_n - \lambda P_n)^{-1} = - \overrightarrow{\prod_{1 \leq k \leq n}} (I + (\lambda - \lambda_k(A))^{-1} A\Delta P_k) P_n,$$

because $A_n P_k = AP_k$ for $k \leq n$. On the other hand,

$$R_\lambda(A)P_n = (A_n - \lambda P_n)^{-1}.$$

But A is compact and therefore A_n tends to A in the operator norm as n tends to ∞. Besides, $\sigma(A_n) \subset \sigma(A)$ and

$$(A_n - \lambda P_n)^{-1} P_n \to (A - \lambda I)^{-1}$$

in the operator norm for any regular λ. We arrive at the result.$\square$

Let A be a normal operator. Then the representation

$$A = \sum_{k=1}^{\infty} \lambda_k(A)\Delta P_k$$

is true (see for instance Dunford and Schwartz (1963)). Hence,

$$A\Delta P_k = \lambda_k(A)\Delta P_k.$$

Since

$$\Delta P_k \Delta P_j = 0 \text{ for } j \neq k,$$

Lemma 2.8.1 gives the equality

$$\lambda R_\lambda(A) = -\sum_{k=1}^{\infty}(I + (\lambda - \lambda_k(A))^{-1}\lambda_k(A)\Delta P_k).$$

But

$$I = \sum_{k=1}^{\infty}\Delta P_k.$$

Thus ,

$$\lambda R_\lambda(A) = -\sum_{k=1}^{\infty}[1 + (\lambda - \lambda_k(A))^{-1}\lambda_k(A)]\Delta P_k =$$

$$-\sum_{k=1}^{\infty}\lambda\Delta P_k(\lambda - \lambda_k(A))^{-1}.$$

Or

$$R_\lambda(A) = \sum_{k=1}^{\infty}(\lambda_k(A) - \lambda)^{-1}\Delta P_k.$$

I.e. we have obtained the well-known result (see e.g. Dunford and Schwartz (1963)).

Thus, Lemma 2.8.1 generalizes the well-known spectral representation for the resolvent of a normal completely continuous operator.

2.8.2   Multiplicative integrals

For a bounded operator-valued function F defined on a real segment $[a, b]$ we define *the right multiplicative integral* as the limit in the uniform operator topology of the sequence of the products

$$\overrightarrow{\prod_{1 \leq k \leq n}} (1 + \Delta F(t_k^{(n)})) \doteq (1 + \Delta F(t_1^{(n)}))(I + \Delta F(t_2^{(n)}))...(I + \Delta F(t_n^{(n)}))$$

as $\max_k |t_k^{(n)} - t_{k-1}^{(n)}|$ tends to zero.

Here

$$\Delta F(t_k^{(n)}) = F(t_k^{(n)}) - F(t_{k-1}^{(n)}), \quad \text{for } k = 1, ..., n$$

and

$$a = t_0^{(n)} < t_1^{(n)} < ... < t_n^{(n)} = b.$$

I.e. the arrow over the symbol of the product means that the indexes of the co-factors increase from left to right.

The right multiplicative integral we denote by

$$\int_{[a,b]}^{\longrightarrow} (1 + dF(t)).$$

For more details about the multiplicative integral see (Gohberg and Krein, 1970), (Brodskii, 1971), (Feintuch and Saeks, 1982).

Let $P(t)$ be an orthogonal resolution of the identity defined on $[a, b]$ and A be a compact linear operator. Then the right multiplicative integral

$$\int_{[a,b]}^{\longrightarrow} (I + AdP(t))$$

is the limit in the uniform operator topology of the sequence of the products

$$\overrightarrow{\prod_{1 \le k \le n}} (I + A\Delta P(t_k)),$$

where

$$\Delta P(t_k) = P(t_k^{(n)}) - P(t_{k-1}^{(n)}), \quad \text{for } k = 1, ..., n.$$

2.8.3   A spectral resolution of a completely continuous operator

A family $E(t)$ of orthogonal projectors defined on a segment of the real axis is called *an orthogonal resolution of the identity* if the following relations are satisfied:

$$E(a) = 0, E(b) = I;$$

and

$$E(t)E(s) = E(min\{t, s\}) \text{ for all } t, s \in [a, b].$$

Definition 2.8.2 *An orthogonal resolution of the identity $P(t)$ $(a \leq t \leq b)$ is said to be a maximal resolution of the identity if any gap $P(t+0)-P(t-0)$ of $P(t)$ (if it exists) is one-dimensional.*

Let $P(t)$ be a continuous on the left maximal resolution of the identity defined on a real segment $[a,b]$. Then we will call it the spectral resolution of an operator A if

$$P(t)AP(t) = AP(t) \text{ for all } t \in [a, b]. \qquad (8.3)$$

As is well known (Gohberg and Krein, 1970), (Brodskii, 1971), any compact linear operator possesses a spectral resolution.

Definition 2.8.3 *Let ΔP be a gap of the spectral resolution $P(t)$ of an operator A. If for some eigenvalue $\lambda(A)$ of A*

$$\Delta PA\Delta P = \lambda(A)\Delta P, \qquad (8.4)$$

then we will call ΔP the gap of $P(t)$ corresponding to $\lambda(A)$.

2.8.4 Statement of the result

Theorem 2.8.4 *Let A be a linear compact operator, and ΔP_k, $k = 1, 2, ...$ be the gaps of its spectral resolution corresponding to eigenvalues $\lambda_k(A)$. Let Q_c be the projector onto the closed linear span of all the root vectors of A. Then for all regular λ*

$$\lambda R_\lambda(A) = - \overrightarrow{\prod_{1 \leq k \leq \infty}} (I + \frac{A_1 \Delta P_k}{\lambda - \lambda_k(A)}) \overrightarrow{\int_{[a,b]}} (I + \lambda^{-1} V_0 dP(t)),$$

where $A_1 = Q_c A Q_c$, $V_0 = A(I - Q_c)$ and $P(t)$ $(a \leq t \leq b)$ is the spectral resolution of V_0.

The proof of this theorem is given in the next subsection.

Notice that this theorem generalizes Lemma 2.8.1 and therefore generalizes the spectral representation of the resolvent of a normal compact operator.

2.8.5   Proof of Theorem 2.8.4

Lemma 2.8.5 *Let V be a Volterra operator with a spectral resolution $P(t)$, then the sequence of the operators*

$$V_n = \sum_{k=1}^{n} P(t_{k-1}^{(n)}) V \Delta P(t_k^{(n)}),$$

$$\Delta P(t_k^{(n)}) = P(t_k^{(n)}) - P(t_{k-1}^{(n)})$$

tends to V in the uniform operator topology as $n \to \infty$.

Proof: We have

$$V - V_n = \sum_{k=1}^{n} \Delta P(t_k^{(n)}) V \Delta P(t_k^{(n)}).$$

Thanks to the well known Lemma 1.5.1 (Gohberg and Krein, 1970) the sequence $\{\|V - V_n\|\}$ tends to zero as n tends to infinity. As claimed. $\square$

Lemma 2.8.6 *Let the $\{P_k\}_{k=0}^{n}$ be an increasing chain of orthoprojectors in H. That is,*

$$0 = range P_0 \subset range P_1 \subset \ldots \subset range P_n = I.$$

Suppose a linear operator W satisfies the relationship

$$P_{k-1} W P_k = W P_k \ for \ k = 1, 2, \ldots, n.$$

Then W is a nilpotent operator and

$$(I - W)^{-1} = \overrightarrow{\prod_{2 \leq k \leq n}} (I + W \Delta P_k)$$

with $\Delta P_k = P_k - P_{k-1}$.

Proof: Set $W_k = W \Delta P_k$. Taking into account that $P_n = I$ and $W P_1 = 0$, we can write down

$$W^j = W P_{n-j+1} W P_{n-j+2} \ldots W P_n = \sum_{2 \leq k_1 < k_2 < \ldots < k_j \leq n} W_{k_1} W_{k_2} .. W_{k_j} \quad (8.5)$$

Hence

$$W^n = 0, \tag{8.6}$$

i.e W is a nilpotent operator.

Further, we have

$$\overrightarrow{\prod_{2 \le k \le n}} (I + W_k) = I + \sum_{k=2}^{n} W_k + \sum_{2 \le k_1 < k_2 \le n} W_{k_1} W_{k_2} + \cdots$$

$$+ W_2 W_3 \ldots W_n.$$

According to (8.5)

$$\overrightarrow{\prod_{2 \le k \le n}} (I + W_k) = \sum_{j=0}^{n-1} W^j.$$

But thanks to (8.6) it follows that

$$(I - W)^{-1} = \sum_{j=0}^{n-1} W^j.$$

This gives the result. $\square$

Lemma 2.8.7 *Let V be a Volterra operator with a spectral resolution $P(t)$ defined on a segment $[a,b]$. Then*

$$(I - V)^{-1} = \int_{[a,b]}^{\rightarrow} (1 + V dP(t))$$

Indeed, this result is due to Lemma 2.8.5 and Lemma 2.8.6.

Lemma 2.8.8 *Let P be a projector onto an invariant subspace of a bounded linear operator A. Then for a regular λ, the equality*

$$\lambda R_\lambda(A) = -(I - APR_\lambda(A)P)(I - A(I - P)R_\lambda(A)(I - P))$$

is true.

Proof: Denote $E = I - P$. Since

$$A = (E + P)A(E + P) \text{ and } EAP = 0,$$

we have

$$A = PAE + PAP + EAE. \tag{8.7}$$

Now, we check the equality

$$R_\lambda(A) = \Psi(\lambda), \tag{8.8}$$

where

$$\Psi(\lambda) \equiv PR_\lambda(A)P - PR_\lambda(A)PAER_\lambda(A)E + ER_\lambda(A)E.$$

In fact, multiplying this equality from the left by $A - I\lambda$ and taking into account the equalities (8.7), $AP = PAP$ and $PAE = 0$, we obtain the relation

$$((A - I\lambda)P + (A - I\lambda)E + PAE)(PR_\lambda(A)P-$$

$$PR_\lambda(A)PAER_\lambda(A)E + ER_\lambda(A)E) =$$

$$P - PAER_\lambda(A)E + E + PAER_\lambda(A)E = I.$$

Similarly, multiplying (8.8) by $A-I\lambda$ from the right and taking into account (8.7), we obtain I. Therefore, (8.8) is correct.

Due to (8.8)

$$I - AR_\lambda(A) = (I - AR_\lambda(A)P)(I - AER_\lambda(A)E).$$

It remains to use the identity

$$I - AR_\lambda(A) = -\lambda R_\lambda(A). \tag{8.9}$$

As claimed.□

Proof of Theorem 2.8.4 : Since Q_c projects onto the invariant subspace of A, due to Lemma 2.8.8 we have

$$\lambda R_\lambda(A) = -(I - AR_\lambda(A)Q_c)(I - AQ_A R_\lambda(A)Q_A)$$

with $Q_A = I - Q_c$. Take into account that $A_1 = AQ_c$ and $R_\lambda(A)Q_c = Q_c R_\lambda(A_1)$. Then

$$\lambda R_\lambda(A) = -(I - A_1 R_\lambda(A_1))(I - AQ_A R_\lambda(A)Q_A). \tag{8.10}$$

For a sufficiently large λ we have

$$R_\lambda(A) = -\sum_{k=0}^{\infty} \frac{A^k}{\lambda^{k+1}}.$$

Notice that $Q_A A = Q_A A Q_A$, because $Q_c = I - Q_A$ projects onto the invariant subspace. Consequently,

$$Q_A A^2 Q_A = Q_A A Q_A A Q_A = Q_A (A Q_A)^2.$$

Similarly,

$$Q_A A^n Q_A = Q_A (A Q_A)^n, \ n = 3, \dots \ .$$

We arrive at the equality

$$Q_A R_\lambda(A) Q_A = -Q_A \sum_{k=0}^{\infty} \frac{(A Q_A)^k}{\lambda^{k+1}} = Q_A R_\lambda(A Q_A)$$

for sufficiently large λ.

Lemma 2.2.4 asserts that $V_0 = A Q_A$ is a Volterra operator. Therefore, we can extend the latter series to all non-zero λ.

Thus, we have due to (8.9)

$$I - A Q_A R_\lambda(A) Q_A = I - V_0 R_\lambda(V_0) = -\lambda R_\lambda(V_0) =$$

$$(I - \lambda^{-1} V_0)^{-1}.$$

Thanks to Lemma 2.8.7

$$(I - \lambda^{-1} V_0)^{-1} = \int_{[a,b]}^{\longrightarrow} (1 + \lambda^{-1} V_0 dP(t)).$$

Equalities (8.9) and (8.10) imply

$$\lambda R_\lambda(A) = \lambda R_\lambda(A_1) \int_{[a,b]}^{\longrightarrow} (1 + \lambda^{-1} V_0 dP(t)).$$

Now we have the result due to Lemma 2.8.1. $\square$

2.9   A relation for the determinant
of a compact operator

We use the function

$$\chi(z) = \max\{|z|, 1\}.$$

of a complex argument z, again.

Theorem 2.9.1 *Let A be a Hilbert-Schmidt operator with a finite trace, and let λ be a complex number such that λ^{-1} does not belong to the spectrum of A. Then the inequality*

$$\|(I - \lambda A)^{-1}det(1 - A\lambda)\| \leq \omega(I - \lambda A)G(\lambda A) \qquad (9.1)$$

is true, where

$$\omega(I - \lambda A) = \prod_{j=1}^{\infty} \chi(1 - \lambda_j(A)\lambda)$$

and

$$G(\lambda A) = \sum_{k=0}^{\infty} |\lambda|^k \frac{g^k(A)}{\sqrt{k!}}.$$

The proof of this theorem is given below in this section.

Corollary 2.9.2 *Under the assumptions of Theorem 2.9.1 the inequality*

$$|det(I - A\lambda)| \leq \|I - A\lambda\| \, \omega(I - A\lambda) \, G(\lambda A) \qquad (9.2)$$

holds.

Notice that if A is a normal operator, then $g(A) = 0$ and therefore $G(A\lambda) = 1$.

Theorem 2.9.1 is exact. In fact, let A be a negative selfadjoint Hilbert-Schmidt operator. Then we can write down

$$\max_{k=1,2,\dots} \lambda_k(A) = 0, \; g(A) = 0 \text{ and } G(A) = 1.$$

Theorem 2.9.1 gives the relation

$$\|(I - A)^{-1}det(I - A)\| \leq \prod_{k=1}^{n}(1 - \lambda_k(A)) = det(I - A),$$

because

$$1 - \lambda_k(A) \geq 1 \text{ for all } k = 1, 2, \ldots,$$

and, consequently,

$$\omega(I - A) = det(I - A)$$

in this case. But A is selfadjoint and thus,

$$\|(I - A)^{-1}\| = \frac{1}{min_k(1 - \lambda_k(A))} = 1$$

I.e., (9.2) is attained.

Corollary 2.9.3 *Under the assumptions of Theorem 2.9.1 the estimate*

$$\|(I - A\lambda)^{-1}\| \leq \frac{G(A\lambda)}{\prod_{k=1}^{\infty} \psi(1 - \lambda\lambda_k(A))} \tag{9.3}$$

is true where

$$\psi(z) = min\{1, |z|\} \text{ for a complex number } z.$$

In fact, we have the equality

$$\chi(z)/|z| = 1/\psi(z)$$

which gives

$$\chi(1 - \lambda\lambda_k(A))/|1 - \lambda\lambda_k(A)| = 1/\psi(1 - \lambda\lambda_k(A)).$$

From this and Theorem 2.9.1 the relation (9.3) follows.

In order to prove Theorem 2.9.1, we will prove a few lemmas.

Lemma 2.9.4 *Let A be a linear operator in a separable Hilbert space and $\{d_k\}$ be an orthonormed basis in H. Then for each $x \in H$*

$$\|Ax\| \leq \| |A||x| \|$$

where $|A|$ is the matrix whose elements with respect to $\{d_k\}$ are the moduli of the elements of A, with respect to $\{d_k\}$, and $|x|$ is the vector whose coordinates are the moduli of the coordinates of the vector x in the basis $\{d_k\}$.

I.e.

$$|x| = \sum_{k=1}^{\infty} |x_k| d_k \text{ with } x_k = (x, d_k).$$

The proof is the evident application of the equality

$$\|Ax\|^2 = \sum_{j=1}^{\infty} |(Ax)_j|^2,$$

where $(.)_j$ means the j-th coordinate. $\square$

Lemma 2.9.5 *Let A be a Hilbert-Schmidt operator with a finite trace and let its root system be complete in H. Let $I - A$ be invertible. Then the following relation is true:*

$$\|(I - A)^{-1} det(I - A)\| \leq \omega(I - A)\|(I - |V|)^{-1}\|,$$

where $|V|$ is the operator whose elements in the Schur basis $\{e_k\}$ are moduli of the nilpotent part V of A in this basis.

That is, the elements of $|V|$ are

$$(|V|e_k, e_j) = |(Ae_k, e_j)| \text{ for } j < k$$

and

$$(|V|e_k, e_j) = 0 \text{ for } j \geq k.$$

Proof: Due to Lemma 2.8.1 we can write

$$(I - A)^{-1} = \overrightarrow{\prod_{1 \leq k \leq \infty}} (I + (1 - \lambda_k(A))^{-1} A \Delta P_k).$$

That is,

$$(I - A)^{-1} = [det(I - A)]^{-1} \overrightarrow{\prod_{1 \leq k \leq \infty}} (I(1 - \lambda_k(A)) + A \Delta P_k). \qquad (9.4)$$

Set

$$K = (I - A)^{-1} det(I - A) \qquad (9.5)$$

and

$$\overline{P}(k) = I - \Delta P_k.$$

Due to the equality

$$A\Delta P_k = \lambda_k(A)\Delta P_k + V\Delta P_k$$

and (9.4) we have

$$K = \overrightarrow{\prod_{1 \leq k \leq \infty}} (I - \lambda_k(A)\overline{P}(k) + V\Delta P_k).$$

Take into account that

$$(I + V\Delta P_k)(I - \lambda_k(A)\overline{P}(k)) = I - \lambda_k(A)\overline{P}(k) + V\Delta P_k,$$

then

$$K = \overrightarrow{\prod_{1 \leq k \leq \infty}} (I + V\Delta P_k)(I - \lambda_k(A)\overline{P}(k)).$$

Put

$$K_j = \overrightarrow{\prod_{j \leq k \leq \infty}} (I + V\Delta P_k)(I - \lambda_k(A)\overline{P}(k)).$$

Consequently,

$$K_1 = K = (I + V\Delta P_1)(I - \lambda_1(A)\overline{P}(1))K_2.$$

It is obvious that

$$|(I - \lambda_k(A)\overline{P}(k))x| \leq (|1 - \lambda_k(A)|\overline{P}(k) + \Delta P_k)|x|,$$

where the inequality is understood for each coordinate with respect to the Schur basis, and the symbol $| \cdot |$ again means that we take moduli of elements of the operators in the Schur basis.

Employing Lemma 2.9.4, we have

$$\|Kx\| \leq \| (I + |V|\Delta P(1))(|1 - \lambda_1(A)|\overline{P}(1) + \Delta P(1))|K_2 x| \|.$$

Further, it is clear that

$$(|1 - \lambda_k(A)|)\overline{P}(k) + \Delta P_k |x| \leq \chi_k |x|,$$

where $\chi_k = \chi(1 - \lambda_k(A))$. Thus,

$$\|Kx\| \leq \chi_1 \| (I + |V|\Delta P_1)|K_2 x| \|.$$

Similarly, due to equality

$$K_2 = (I + V\Delta P_2)(I - \lambda_2(A)\overline{P}(2))K_3$$

it can be written

$$\|Kx\| \leq \chi_1 \chi_2 \|(I + |V|\Delta P(1))(I + |V|\Delta P(2))|K_3 x| \|.$$

Repeating these arguments, we obtain:

$$\|Kx\| \leq \prod_{1 \leq j \leq n} \chi_j \| \overrightarrow{\prod_{1 \leq k \leq n}} (I + |V|\Delta P_k)|K_{n+1} x| \|.$$

By Lemma 2.8.1 the sequence of operators K_n tends to I as n tends to infinity. Thus,

$$\|Kx\| \leq \prod_{1 \leq j \leq \infty} \chi_j \| \overrightarrow{\prod_{1 \leq k \leq \infty}} (I + |V|\Delta P_k)|x| \|.$$

According to Lemma 2.8.1

$$\overrightarrow{\prod_{1 \leq k \leq \infty}} (I + |V|\Delta P_k) \equiv (I - |V|)^{-1}.$$

This and (9.5) give the result.$\square$

Proof of Theorem 2.9.1: Employing Corollary 2.4.7, we obtain

$$\|(I - |V|)^{-1}\| \leq \sum_{k=0}^{\infty} \frac{N^k(|V|)}{(k!)^{1/2}}.$$

Taking into account that $N(V) = N(|V|)$ and using Lemma 2.4.6, we arrive at the inequality

$$\|(I - |V|)^{-1}\| \leq \sum_{k=0}^{\infty} g^k(A)(k!)^{-1/2} = G(A)$$

Substituting this inequality into the assertion of Lemma 2.9.5 and replacing A by λA, we conclude that

$$\|(I - \lambda A)^{-1} det(I - \lambda A)\| \le G(A\lambda) \prod_{j=1}^{\infty} \chi(1 - \lambda_j(A)\lambda). \qquad (9.6)$$

I.e. in the case of completeness of the root vectors the theorem is proved.

Let us consider now the general case. Thanks to Corollary 2.4.5 there is a sequence $\{A_n\}$ of compact operators having finite dimensional ranges such that $g(A_n) \to g(A)$ as $n \to \infty$. This relation implies that

$$\lim_{n \to \infty} G(\lambda(A_n)) = G(\lambda(A)). \qquad (9.7)$$

Due to the well-known Corollary 4.1.1 (Gohberg and Krein, 1969) we have

$$|det(I - \lambda A_n) - det(I - \lambda A)| \to 0 \text{ as } n \to \infty.$$

Moreover, Lemma 2.4.2 asserts that for any neighborhood M of $\sigma(A)$ there is such n_0 that $\sigma(A_n)$ belongs to M for all $n \ge n_0$ and

$$R_\mu(A) - R_\mu(A_n) \to 0 \text{ as } n \to \infty$$

in the operator norm for any regular μ not belonging to M. Therefore,

$$(I - \lambda A_n)^{-1} \to (I - \lambda A)^{-1} \text{ as } n \to \infty$$

in the operator norm.

Above we have proved the theorem for an operator whose system of the root vectors is complete in H, and in particular, for finite dimensional operators. Consequently, the assertion now follows from (9.6) and (9.7). The proof is complete. $\square$

2.10 Nilpotent parts of Neumann-Schatten operators

Let A be a Hilbert-Schmidt operator. Due to Lemma 2.4.6 we can write down the equality

$$N_2^2(V) = N_2^2(A) - N_2^2(S) = g^2(A), \qquad (10.1)$$

where S is the diagonal part, and V is the nilpotent one of the operator A. In this section we will derive analogous results for operators from the Neumann-Schatten ideals C_p with $p = 4$ and $p = 6$.

At the beginning we consider the case $p = 4$.

Lemma 2.10.1 *Let $A \in C_4$. Then the relation*

$$N_2(V^2) \leq \sqrt{N_4^4(A) - N_4^4(S)} + 2N_4(S)\sqrt{N_4^2(A) - N_4^2(S)} \qquad (10.2)$$

is true.

Proof: The triangular representation (2.1) implies

$$A^2 = S^2 + VS + SV + V^2.$$

We invoke Lemma 2.6.2. It asserts that

$$SV + VS + V^2 \qquad (10.3)$$

is a Volterra operator and the Schur basis of this operator coincides with the Schur basis of A. But the Schur basis of A coincides with the Schur basis of the operator A^2, and (10.3) is the nilpotent part of the operator A^2. According to (10.1)

$$N_2^2(SV + VS + V^2) = N_2^2(A^2) - N_2^2(S^2). \qquad (10.4)$$

Further, it is simple to check that VS is the nilpotent part of the operator AS. Therefore, (10.1) gives

$$N_2^2(AS) = N_2^2(S^2) + N_2^2(VS). \qquad (10.5)$$

Similarly, the operator SV is the nilpotent part of the operator SA. Thus, we can write down

$$N_2^2(SA) = N_2^2(S^2) + N_2^2(SV). \qquad (10.6)$$

Taking into account that

$$N_2(SV + VS) \leq N_2(SV) + N_2(VS),$$

we obtain

$$N_2(SV + VS) \leq \sqrt{N_2^2(SA) - N_2^2(S^2)} + \sqrt{N_2^2(AS) - N_2^2(S^2)}.$$

Now, (10.4) implies

$$N_2(V^2) \leq N_2(SV + VS) + \sqrt{N_2^2(A^2) - N_2^2(S^2)}.$$

Due to (10.5) and (10.6)

$$N_2(V^2) \leq \sqrt{N_2^2(A^2) - N_2^2(S^2)} + \sqrt{N_2^2(SA) - N_2^2(S^2)} +$$

$$\sqrt{N_2^2(AS) - N_2^2(S^2)}. \tag{10.7}$$

But $N_2^2(A^2) \leq N_4^4(A)$ and $N_2^2(S^2) = N_4^4(S)$. Exploiting (1.6), we have

$$N_2(AS) \leq N_4(S)N_4(A)$$

and

$$N_2(SA) \leq N_4(S)N_4(A).$$

Thus, (10.7) implies (10.2). As claimed.□

Corollary 2.10.2 *Let $A \in C_4$. Then the relation*

$$N_2(V^2) \leq (1 + \sqrt{2})\sqrt{N_4^4(A) - N_4^4(S)} \tag{10.8}$$

is true.

Indeed, we have

$$2N_4^2(A)N_2^2(S) \leq N_4^4(A) + N_4^4(S).$$

Hence,

$$2N_4(S)\sqrt{N_4^2(A) - N_4^2(S)} = \sqrt{4(N_4^2(A)N_4^2(S) - N_4^4(S))} \leq$$

$$\sqrt{2(N_4^4(A) - N_4^4(S))}.$$

Now the result is due to Lemma 2.10.1.

The inequality (10.8) and therefore, Lemma 2.10.1 are precise: if A is normal then $N_4(A) = N_4(S)$ and (10.8) ensures the equality $V = 0$.

Lemma 2.10.3 *Let $A \in C_6$. Then the relation*

$$N_2(V^3) \leq \sqrt{N_6^6(A) - N_6^6(S)} + 2N_6(S)\sqrt{N_6^4(A) - N_6^4(S)}$$

is true.

Proof: The triangular representation (2.1) implies

$$A^3 = S^3 + VS^2 + SVS + S^2V + V^2S + VSV + SV^2 + V^3.$$

Put

$$V_3 = VS^2 + SVS + S^2V + V^2S + VSV + SV^2$$

and use Lemma 2.6.2. It asserts that $V_3 + V^3$ is a Volterra operator and the Schur basis of this operator coincides with the Schur basis of A. Therefore, it coincides with the Schur basis of the operator A^3.

Since $A^3 - V^3 - V_3 = S^3$ is a normal operator, $V_3 + V^3$ is the nilpotent part of A^3. According to (10.1)

$$N_2^2(V_3 + V^3) = N_2^2(A^3) - N_2^2(S^3). \qquad (10.9)$$

Thanks to the triangular representation (2.1), we can write down the relation

$$A^2S = (S^2 + VS + SV + V^2)S =$$
$$S^3 + VS^2 + SVS + V^2S$$

Repeating our reasoning above we can assert that

$$VS^2 + SVS + V^2S$$

is the nilpotent part of the operator A^2S. Using (10.1) we easily obtain

$$N_2^2(VS^2 + SVS + V^2S) = N_2^2(A^2S) - N_2^2(S^3). \qquad (10.10)$$

Besides,

$$ASA = S^3 + S^2V + VS^2V + VSV.$$

So, the operator

$$S^2V + VS^2 + VSV$$

is the nilpotent part of the operator ASA. Exploiting (10.1) again, we arrive at the equality

$$N_2^2(ASA) = N_2^2(S^3) + N_2^2(S^2V + VS^2 + VSV) \qquad (10.11)$$

On the other hand, (10.9) implies

$$N_2^2(V^3) \le N_2^2(V_3) + \sqrt{N_2^2(A^3) - N_2^2(S^3)} \le$$

$$N(VS^2 + SVS + V^2S) + N(S^2V + VSV + SV^2) +$$

$$\sqrt{N_2^2(A^3) - N_2^2(S^3)}$$

This, (10.10) and (10.11) give

$$N^2(V^3) \le \sqrt{N_2^2(A^3) - N_2^2(S^3)} + \sqrt{N_2^2(A^2S) - N_2^2(S^3)} +$$

$$\sqrt{N^2(ASA) - N_2^2(S^3)}. \qquad (10.12)$$

Take into account that

$$N_2^2(A^3) \le N_6^6(A) \text{ and } N_2^2(S^3) = N_6^6(S).$$

Exploiting (1.6), we have

$$N_2^2(ASA) \le N_6^2(S)N_6^4(A)$$

and

$$N_2^2(A^2S) \le N_6^2(S)N_6^4(A).$$

Thus, (10.12) implies the stated result.□

Corollary 2.10.4 *Let* $A \in C_6$. *Then the relation*

$$N_2(V^3) \le (1 + \frac{2}{\sqrt{3}})\sqrt{N_6^6(A) - N_6^6(S)} \qquad (10.13)$$

is true.

Indeed, we use the inequality

$$3ab^2 \leq a^3 + b^3 \; ; a, b \geq 0.$$

It implies:

$$3N_6^2(S)N_6^4(A) \leq N_6^6(S) + N_6^6(A).$$

Consequently,

$$N_6(S)\sqrt{N_6^4(A) - N_6^4(S)} = \sqrt{N_6^4(A)N_6^2(S) - N_6^6(S)} \leq$$

$$\sqrt{N_6^6(A)/3 - (2/3)N_6^6(S)} \leq \sqrt{N_6^6(A)/3 - N_6^6(S)/3}.$$

Now the result follows from Lemma 2.10.3.

The inequality (10.13) and therefore, Lemma 2.10.3 are precise. Indeed, let $A \in C_6$ be a normal operator. Then $A = S$ and (10.13) ensure the equality $V = 0$.

It is not hard to show that for any natural number $r \geq 1$ there is a constant a_r independent of A (but dependent on r) such that

$$N_2^2(V^r) \leq a_r(N_{2r}^{2r}(A) - N_{2r}^{2r}(S)).$$

2.11   Equalities for eigenvalues of a Hilbert-Schmidt operator

Lemma 2.11.1 *For any $A \in C_2$ the equality*

$$g^2(A) \equiv N^2(A) - \sum_{k=1}^{\infty} |\lambda_k(A)|^2 =$$

$$2N^2(A_I) - 2\sum_{k=1}^{\infty} |\Im \lambda_k(A)|^2$$

holds.

Here as above $A_I = (A - A^*)/2i$.

Proof: Corollary 2.4.5 asserts that there is a sequence $\{A_n\}$ of linear operators having finite ranges such that $N(A_n) \to N(A)$, $g(A_n) \to g(A)$ as $n \to \infty$. By Corollary 1.3.7

$$g^2(A_n) = 2N^2((A_n)_I) - 2 \sum_{k=1}^{n} |\Im\lambda_k(A_n)|^2.$$

Here $(A_n)_I = (A_n - A_n^*)/2i$. Letting $n \to \infty$ we arrive at the result. The details are left to the reader. $\square$

Corollary 2.11.2 *For any $A \in C_2$ and real numbers t, τ the equality*

$$N^2(Ae^{it} - A^* e^{-it}) - \sum_{k=1}^{\infty} |e^{it}\lambda_k - e^{-it}\overline{\lambda}_k|^2 =$$

$$N^2(Ae^{i\tau} - A^* e^{-i\tau}) - \sum_{k=1}^{\infty} |e^{i\tau}\lambda_k - e^{-i\tau}\overline{\lambda}_k|^2 =$$

is true.

The proof consists in replacing A by Ae^{it} and $Ae^{i\tau}$ and using Lemma 2.11.1 (see also the proof of Lemma 1.3.10). In particular, take $t = 0$ and $\tau = \pi/2$. Due to Corollary 2.11.2

$$N^2(A_I) - \sum_{k=1}^{\infty} |\Im\lambda_k(A)|^2 = N^2(A_R) - \sum_{k=1}^{\infty} |\Re\lambda_k(A)|^2$$

with $A_R = (A + A^*)/2$.

2.12 Notes

Sections 2.1 and 2.2 For the exposition of the main notions listed in Sections 2.1 and 2.2 see as above mentioned (Ahiezer and Glazman, 1981), (Dunford and Schwartz, 1963), (Gohberg and Krein, 1969), (Pietsch, 1988).

Sections 2.3-2.5 Theorem 2.4.1 is due to Gil' (1979a). The results presented in Sections 2.4 and 2.5 are adapted from Gil' (1979b) and Gil' (1983a).

Sections 2.6-2.7   Theorem 2.6.1 is due to Gil' (1991). The contents of Section 2.7 are based on Gil' (1979a) and Gil' (1992).

Sections 2.8   Theorem 2.8.1 is announced in Gil' (1973b) and is proved in Gil' (1984a).

Section 2.9   Theorem 2.9.1 and its corollaries are new. They improve the Carleman inequality (Dunford and Schwartz, 1963). Another relation between resolvents and determinants is in Gohberg and Krein (1969, Chapter 5).

The results presented in **Sections 2.10 and 2.11** are new.

Chapter 3

Functions of Nonselfadjoint Operators

This chapter is devoted to functions of noncompact nonselfadjoint operators.

In Section 3.1 some well-known notions of the theory of linear operators in a Hilbert space are listed.

Section 3.2 deals with the so called triangular representation of linear operators. This representation is the basis for obtaining the norm estimations.

Section 3.3 analyzes general operators admitting the triangular representation. An estimation for the norm is obtained.

In Section 3.4 we obtain one of the main results of the chapter. Namely we do the estimation for the resolvent of a quasi-Hermitian operator, i.e. an operator whose imaginary component is a compact operator.

In Section 3.5 we consider the resolvent of a quasiunitary operator, i.e. an operator which is a sum of a unitary and a compact operator.

Section 3.6 is concerned with regular functions of operators admitting the triangular representation.

Regular functions of an operator with Hilbert-Schmidt's imaginary component are analyzed in Section 3.7.

Regular functions of quasiunitary operators are considered in Section 3.8.

A multiplicative representation for a resolvent of a nonselfadjoint operators is derived in Section 3.9.

3.1   Terminology

In this section we recall very briefly some notions of spectral theory of operators acting in Hilbert space. For more details any textbook on Hilbert can be consulted (for instance (Ahiezer and Glazman, 1981), (Dunford and Schwartz, 1963), (Gohberg and Krein, 1969), etc. See also Section 2.1).

3.1.1   Bounded linear operators

Let H be a Hilbert space with a scalar product $(.,.)$ and A be a bounded operator acting in H.

For any bounded selfadjoint operator A there exists an orthogonal resolution of the identity $E(s)$ (see Section 2.8) defined on a segment $[a,b]$ of the real axis such that

$$A = \int_a^b sdE(s). \tag{1.1}$$

Any unitary operator A can be represented as

$$A = \int_0^{2\pi} e^{is}dE(s), \tag{1.2}$$

where $E(s)$ is an orthogonal resolution of the identity defined on $[0, 2\pi]$.

The integrals in (1.1) and (1.2) are the strong limits of the corresponding operator sums.

A linear bounded operator A is said to be *a quasi-Hermitian operator* if

$$A_I \equiv (A - A^*)/2i \text{ is a completely continuous operator.} \tag{1.3}$$

A linear bounded operator A is called *a quasiunitary one* if

$$A^*A - I \text{ is a completely continuous operator .} \qquad (1.4)$$

It is well known (see e.g. Gohberg and Krein (1969), Sections 1.5 and 1.6) that the nonreal spectrum of a quasi-Hermitian operator consists of eigenvalues of a finite multiplicity, and

$$\sum_{k=1}^{n} |\Im \lambda_j(A)|^p \leq \sum_{k=1}^{n} s_j^p(A_I); \ n = 1, 2., \ldots \qquad (1.5)$$

for any $p \geq 1$.

3.1.2 Unbounded operators in a Hilbert space

Let there be a linear manifold $D(A)$ such that the relation $h \in D(A)$ implies $Ah \in H$ for a given linear operator A. Then the set $D(A)$ is called *the domain of* A.

Let $D(A)$ be dense in H. Then the set of vectors g satisfying

$$|(Af, g)| \leq c\|f\| \text{ for all } f \in D(A)$$

with a constant c is the domain of *the adjoint operator* A^*, and besides

$$(Af, g) = (f, A^*g) \text{ for all } f \in D(A) \text{ and } g \in D(A^*).$$

An unbounded operator A is selfadjoint if

$$D(A) = D(A^*) \text{ and } Ah = A^*h$$

for each $h \in D(A)$. An unbounded selfadjoint operator possesses the unbounded real spectrum, and the spectral representation (1.1) is true.

An unbounded operator A is normal if $D(AA^*) = D(A^*A)$ and $AA^*h = A^*Ah$ for each $h \in D(A^*A)$. An unbounded normal operator has an unbounded spectrum.

An unbounded operator A *is quasi-Hermitian if* $D(A) = D(A^*)$ *and the* operator $A - A^*$ has a completely continuous extension.

3.2   Triangular representations

In the sequel H is a separable Hilbert space.

Let an orthogonal resolution of the identity $P(t)$ in H defined on a (finite or infinite) segment $[a, b]$ of the real axis be a maximal resolution of the identity. Recall (see Section 2.8), that this means that $P(t)$ is a family of orthogonal projectors in H, i.e. $P^2(t) = P(t)$ and $P^*(t) = P(t)$, such that

a) $P(a) = 0$, $P(b) = I$,

b) $P(t)P(s) = P(min(t, s))$,

c) $P(t)$ is continuous on the left in the strong topology:

$$\|P(t - 0)h - P(t)h\| = 0 \text{ for any } h \in H,$$

d) every gap $P(t + 0) - P(t)$ (if it exists) is one-dimensional.

A scalar function ϕ is said to be *P-measurable if it is measurable with respect to the scalar measure $(P(t)g, g)$ with an arbitrary $g \in H$.*

Recall (see Section 2.8) that a maximal orthogonal resolution of the identity $P(t)$ defined on a segment $[a, b]$ of the real axis is a spectral resolution of a linear operator A if

$$P(t)AP(t) = AP(t) \text{ for all } t \in [a, b]$$

Definition 3.2.1 *Let a linear operator A in a separable Hilbert space have a spectral resolution $P(t)$ defined on a (finite or infinite) real segment $[a, b]$. Suppose that there are a Volterra operator V, and a P-measurable scalar-valued function $\phi(t)$ such that the equality*

$$A = S + V, \tag{2.1}$$

holds, where

$$Sh = \int_a^b \phi(t)dP(t)h, \ h \in D(A). \tag{2.2}$$

Then we will call (2.1), S and V the triangular representation, the diagonal part and the nilpotent part of A, respectively.

It is assumed that the integral in (2.2) converges strongly for any $h \in D(A)$.

It is clear that (2.2) implies the equality

$$P(t)VP(t) = VP(t) \text{ for all } t \in [a, b]. \tag{2.3}$$

Lemma 3.2.2 *For a linear completely continuous operator V in H, let there be a maximal orthogonal resolution of the identity $P(t)$, $a \leq t \leq b$ satisfying the conditions (2.3) and*

$$(P(t_0 + 0) - P(t_0))V(P(t_0 + 0) - P(t_0)) = 0 \tag{2.4}$$

for every gap $P(t_0 + 0) - P(t_0)$ of $P(t)$. Then V is a Volterra operator.

In particular, if $P(t)$ is continuous in t in the strong topology, then any linear completely continuous operator V satisfying (2.3) is a Volterra one.

Proof: Indeed this result is due to the well-known Corollary 1 to Theorem 17.1 of the book by Brodskii (1971).□

Lemma 3.2.3 *Let V be a Volterra operator in H, and $P(t)$ be a maximal orthogonal resolution of the identity satisfying the equality (2.3). Then the equality (2.4) holds for every gap $P(t_0 + 0) - P(t_0)$ of $P(t)$.*

Proof: This result is due to the wellkown equality (1.3.1) from the book by Gohberg and Krein (1970, p.15) .□

Lemma 3.2.4 *Let V be a Volterra operator with a spectral resolution $P(t)$ defined on a real segment $[a,b]$. Let a bounded operator A satisfy the condition*

$$P(t)AP(t) = AP(t) \text{ for all } t \in [a, b].$$

Then VA and AV are Volterra operators and $P(t)$ is their spectral resolution.

Proof: It is obvious that

$$P(t)VAP(t) = VP(t)AP(t) = VAP(t). \tag{2.5}$$

Let now $Q = P(t_0 + 0) - P(t_0)$ be a gap of $P(t)$. Then according to Lemma 3.2.3 the equality (2.4) holds. Further, we have

$$QVAQ = QVA(P(t_0+0)-P(t_0)) = QV[P(t_0+0)AP(t_0+0)-P(t_0)AP(t_0)] =$$

$$QV[(P(t_0)+Q)A(P(t_0)+Q) - P(t_0)AP(t_0)] = QV[P(t_0)AQ + QAP(t_0)].$$

Since $QP(t_0) = 0$ and $P(t)$ projects onto the invariant subspaces, we obtain $QVAP(t_0) = 0$.

Due to Lemma 3.2.3 this relation and the equality (2.5) imply that VA is a Volterra operator. Similarly we can prove AV is a Volterra one. $\square$

Lemma 3.2.5 *Let bounded linear operators A_1 and A_2 admit the triangular representation with the same spectral resolution $P(t)$. Then the operator $A = A_1 + A_2$ admits the triangular representation with the spectral resolution $P(t)$. Moreover, the diagonal part of A is*

$$S = S_1 + S_2$$

and the nilpotent part of A is

$$V = V_1 + V_2,$$

where S_k and V_k are the diagonal part and the nilpotent one of the operator A_k, $k = 1, 2$, respectively.

Proof: According to (2.2) S_k $(k = 1, 2)$ has the shape

$$S_k = \int_a^b \phi_k(t)dP(t), \tag{2.6}$$

where $\phi_k(t)$ are $P(t)$ measurable functions. Hence the operator $S = S_1 + S_2$ has the form (2.3) with $\phi(t) = \phi_1(t) + \phi_2(t)$.

According to (2.3)

$$P_k(t)VP_k(t) = V_k P(t) \text{ for all } t \in [a, b]. \qquad (2.7)$$

Exploiting Lemma 3.2.4 we can write down

$$(P(t_0 + 0) - P(t_0))V_k(P(t_0 + 0) - P(t_0)) = 0; k = 1, 2 \qquad (2.8)$$

for every gap $P(t_0+0)-P(t_0)$ of $P(t)$. These relations imply the equalities
(2.3) and (2.4). Using Lemma 3.2.3 we can assert that $V = V_1+V_2$ is indeed
a Volterra operator. This proves the result.□

Corollary 3.2.6 *Let V_1, V_2 be Volterra operators with the same spectral
resolution $P(t)$. Then $V = V_1 + V_2$ is a Volterra operator with the spectral
resolution $P(t)$.*

Lemma 3.2.7 *Let bounded linear operators A_1, A_2 admit the triangular
representation with the same spectral resolution $P(t)$ Then the operator
$A = A_1 A_2$ admits the triangular representation with the spectral resolution
$P(t)$. Besides, the diagonal part of A is*

$$S = S_1 S_2$$

and the nilpotent part of A is

$$V = V_1 V_2 + S_1 V_2 + V_1 S_2,$$

*where S_k and V_k are the diagonal part and the nilpotent one of the operator
$A_k, k = 1, 2$, respectively.*

Proof: Because S_k has the form of (2.6), then

$$S \equiv S_1 S_2 = \int_a^b \phi_1(t)dP(t). \int_a^b \phi_2(s)dP(s).$$

But

$$dP(t)dP(s) = 0 \text{ for } t \neq s \text{ and } dP(t)dP(t) = dP(t).$$

Thus we obtain the relation (2.2) with $\phi(s) = \phi_1(s)\phi_2(s)$. According to (2.1) we can write down $A_k = S_k + V_k$. Thus,

$$A = A_1 A_2 = S_1 S_2 + S_1 V_2 + V_1 S_2 + V_1 V_2.$$

Due to Lemma 3.2.4 the operators $S_1 V_2$, $V_1 S_2$ and $V_1 V_2$ are Volterra ones because all of them have common invariant subspaces. It remains to use Corollary 3.2.6.

Lemma 3.2.5 and Lemma 3.2.7 give the following:

Corollary 3.2.8 *The set of all bounded operators admitting the triangular representation with the same spectral resolution is a ring.*

Lemma 3.2.9 *Let A be a bounded linear operator admitting the triangular representation with a spectral resolution $P(t)$. Let V and S be the nilpotent and diagonal parts of A, respectively. Then the operators S^*V and SV are Volterra operators with the same spectral resolution $P(t)$.*

Proof: It is clear that S^*V satisfies the condition

$$P(t)S^* = S^*P(t) \text{ for all } t \in [a, b],$$

where $P(t)$ is the spectral resolution of V. Now Lemma 3.2.4 ensures the result. □

Corollary 3.2.10 *Let A admit the triangular representation (2.1) and let V and S be its nilpotent and diagonal parts, respectively. Then the equalities*

$$Tr(SV) = Tr(S^*V) = 0$$

are true.

Lemma 3.2.11 *Let A be a (generally unbounded) linear operator admitting the triangular representation with a spectral resolution P. Let V and S be the nilpotent and diagonal parts of A, respectively. Then for any regular point λ of S, the operator $VR_\lambda(S)$ is Volterra one with the same spectral resolution.*

Proof: It is clear that

$$P(t)R_\lambda(S) = R_\lambda(S)P(t) \text{ for all } t \in [a, b].$$

Now Lemma 3.2.4 ensures the result. □

Lemma 3.2.12 *Let a linear operator A admit the triangular representation, and let S be its diagonal part. Then $\sigma(A) = \sigma(S)$.*

Proof: Let λ be a regular point of the operator S. According to the representation (2.1) we obtain

$$R_\lambda(A) = (S + V - \lambda I)^{-1} = R_\lambda(S)(I + V R_\lambda(S))^{-1}.$$

Operator $V R_\lambda(S)$ for a regular point λ of the operator S is a Volterra one due to Lemma 3.2.11. Hence,

$$(I + V R_\lambda(S))^{-1} = \sum_{k=0}^{\infty}(V R_\lambda(S))^k(-1)^k \qquad (2.9)$$

and the series converges in the norm. Thus

$$R_\lambda(A) = R_\lambda(S)\sum_{k=0}^{\infty}(V R_\lambda(S))^k(-1)^k. \qquad (2.10)$$

From this it follows that λ is the regular point of A.

Let now $\lambda \in \sigma(S)$. Then (2.10) implies that $\lambda \in \sigma(A)$. These relations prove the assertion. □

3.3 Preliminary estimations for resolvents

In this section we establish a norm estimation for the resolvent of operators admitting the triangular representation.

Lemma 3.3.1 *Let A admit the triangular representation (2.1) and let V be its nilpotent part. If*

$$V \in C_{2r} \text{ for some positive integer } r, \qquad (3.1)$$

then the inequality

$$\|R_\lambda(A)\| \le \sum_{m=0}^{r-1} \sum_{k=0}^{\infty} \frac{N_{2r}^{kr+m}(V)}{\rho^{rk+m+1}(A,\lambda)\sqrt{k!}} \tag{3.2}$$

is true for all regular λ .

Proof: 1. Denote

$$B_\lambda = -VR_\lambda(S)$$

and once more apply the identity

$$(I - B_\lambda)(I + B_\lambda + \ldots + B_\lambda^{r-1}) = I - B_\lambda^r.$$

From this it follows that

$$(I - B_\lambda)^{-1} = (I + B_\lambda + \ldots + B_\lambda^{r-1})(I - B_\lambda^r)^{-1}.$$

Due to Lemma 3.2.11 B_λ is a Volterra operator. Consequently, the series

$$(I - B_\lambda^r)^{-1} = \sum_{k=0}^{\infty} B_\lambda^{kr}$$

converges in the operator norm for all regular points of S. Thus,

$$(I - B_\lambda)^{-1} = (I + B_\lambda + \ldots + B_\lambda^{r-1})(I - B_\lambda^r)^{-1} = \sum_{k=0}^{\infty} B_\lambda^{kr}. \tag{3.3}$$

According to (2.1) we can write down

$$R_\lambda(A) = R_\lambda(S)(I - B_\lambda)^{-1}.$$

Therefore due to (3.3)

$$R_\lambda(A) = R_\lambda(S) = \sum_{m=0}^{r-1} \sum_{k=0}^{\infty} B_\lambda^{kr+m}. \tag{3.4}$$

2. By Lemma 3.2.12 $\sigma(A) = \sigma(S)$, and therefore, $R_\lambda(S)$ is a bounded operator for all regular points λ of A. Hence, $B_\lambda \in C_{2r}$. This implies that B^r is a Hilbert-Schmidt operator, because

$$N_2(B_\lambda^r) \equiv N(B_\lambda^r) \le N_{2r}^r(B_\lambda) \tag{3.5}$$

(see Section 2.1). Now we use the estimate for powers of a Volterra operator of Hilbert-Schmidt's kind obtained in Section 2.3. It gives

$$\|B_\lambda^{rk}\| \le \frac{N^k(B_\lambda^r)}{\sqrt{k!}}.$$

The relation (3.5) implies

$$\|B_\lambda^{rk}\| \le \frac{N_{2r}^{kr}(B_\lambda)}{\sqrt{k!}}.$$

It is clear that

$$N_{2r}(B_\lambda) = N_{2r}(V R_\lambda(S)) \le N_{2r}(V)\|R_\lambda(S)\|.$$

Since S is normal and $\sigma(A) = \sigma(S)$,

$$\|R_\lambda(S)\| = \rho^{-1}(S, \lambda) = \rho^{-1}(A, \lambda). \qquad (3.6)$$

Hence,

$$N_{2r}(B_\lambda) \le \frac{N_{2r}(V)}{\rho(A, \lambda)}. \qquad (3.7)$$

Thus,

$$\|B_\lambda^{rk}\| \le \frac{N_{2r}^{kr}(V)}{\rho^{kr}(A, \lambda)\sqrt{k!}}.$$

Evidently,

$$\|B_\lambda^m\| \le N_{2r}^m(B_\lambda).$$

The relation (3.7) implies

$$\|B_\lambda^m\| \le \frac{N_{2r}^m(V)}{\rho^m(A, \lambda)}.$$

Therefore,

$$\|B_\lambda^{rk+m}\| \le \frac{N_{2r}^{kr+m}(V)}{\rho^{kr+m}(A, \lambda)\sqrt{k!}}.$$

From this, taking into account (3.4), we can write down

$$\|R_\lambda(A)\| \le \|R_\lambda(S)\| \sum_{m=0}^{r-1} \sum_{k=0}^{\infty} \frac{N_{2r}^{kr+m}(V)}{\rho^{kr+m}(A, \lambda)\sqrt{k!}}.$$

Now, the relation (3.6) ensures the result.□

3.4   Resolvents of quasi-Hermitian operators

3.4.1   Representations of quasi-Hermitian operators

Denote by C_ω the Macaev ideal. Recall (Gohberg and Krein, 1970) that
it is the set of compact operators K in a separable Hilbert space with the
property

$$\sum_{k=1}^{\infty} \frac{s_k(K)}{2k+1} < \infty.$$

Theorem 3.4.1 *Let A be a generally unbounded quasi-Hermitian operator
satisfying the condition*

$$A - A^* \in C_\omega. \tag{4.1}$$

Then A admits the triangular representation.

Proof: Due to De Branges (1965, p.69) we can assert that there is an orthogonal resolution of the identity $P(x)$, $-\infty \le x \le \infty$ and nondecreasing
functions $h(x)$ such that under (4.1) the representation

$$A = \int_{-\infty}^{\infty} h(t)dP(t) + 2i \int_{-\infty}^{\infty} P(t)A_I dP(t) \tag{4.2}$$

holds.

The second integral in (4.2) is defined in the following way. The infinite
integral is defined as the limit of the finite integrals

$$\int_a^b P(t)A_I dP(t).$$

The finite integrals are defined as the limit of Stieltjes sums

$$L_n = \frac{1}{2} \sum_{k=1}^{n} [P(t_{k-1}) + P(t_k)]A_I \Delta P_k,$$

where

$$t_k = t_k^{(n)}; \ \Delta P_k = P(t_k) - P(t_{k-1}); \ a = t_0 < t_1 \ldots < t_n = b.$$

All the limits are taken in the operator norm.

We can write down

$$L_n = \frac{1}{2}\sum_{k=1}^{n}[2P(t_{k-1}) + \Delta P_k]A_I\Delta P_k = U_n + S_n$$

with

$$U_n = \sum_{k=1}^{n}P(t_{k-1})A_I\Delta P_k$$

and

$$S_n = 1/2\sum_{k=1}^{n}\Delta P_k A_I \Delta P_k. \qquad (4.3)$$

The sequence $\{S_n\}$ converges in the norm due to the well-known Lemma 1.5.1 from (Gohberg and Krein, 1970). We denote its limit by T. Thanks to the well-known Theorem 2.5.1 from (Gohberg and Krein, 1969), each operator S_n belongs to the ideal C_ω. According to the well-known Theorem 3.5.1 from (Gohberg and Krein, 1969), the operator T belongs to the ideal C_ω too. Using (4.3) and taking into account that T is a compact Hermitian operator, we have

$$T = \sum_{k=1}^{\infty}\lambda_k(T)\Delta P_k,$$

where $\lambda(T)$ are all eigenvalues of T taking into account their multiplicities, ΔP_k are the orthogonal projectors onto eigensubspaces of T. Thus, we arrive at (2.2) with

$$S = \int_{-\infty}^{\infty} h(t)dP(t) + 2iT$$

and

$$\phi(t) = h(t) + 2i\psi(t),$$

where $\psi(t)$ is the function whose values coincide with $\lambda_k(T)$.

Further, it is directly checked that U_n is a nilpotent operator: $(U_n)^n = 0$. Besides, the sequence $\{U_n\}$ converges in the operator norm, because the second integral in (4.2) and $\{S_n\}$ converges in this norm. We denote the limit of the the sequence $\{U_n\}$ by U. It is a Volterra operator, since the limit in the operator norm of a sequence of Volterra operators is a Volterra one (see for instance Lemma 2.17.1 from the book by Brodskii (1971)). From this we easily obtain (2.1) and (2.3).

It is well known that any completely continuous linear operator has a maximal chain of invariant subspaces which has one-dimensional gaps only (see Gohberg and Krein (1970) and Brodskii (1971)). This means that we can assume that $P(x)$ has one-dimensional gaps only (if they exist).

The proof is complete. □

3.4.2   Estimates for resolvents of operators with Hilbert-Schmidt's imaginary components

In this subsection we obtain an estimation for the norm of the resolvent of a quasi-Hermitian operator under the following condition:

$$A_I \text{ is a Hilbert-Schmidt operator.} \tag{4.4}$$

Let us introduce the quantity

$$g_1(A) \equiv \sqrt{2}(N^2(A_I) - \sum_{k=1}^{\infty} |\Im\lambda_k(A)|^2)^{1/2}. \tag{4.5}$$

Theorem 3.4.2 *Let a quasi-Hermitian generally unbounded operator A satisfy the condition (4.4), then the inequality*

$$\|R_\lambda(A)\| \leq \sum_{k=0}^{\infty} \frac{g_1^k(A)}{\rho^{k+1}(A,\lambda)\sqrt{k!}} \text{ for all regular } \lambda. \tag{4.6}$$

is true.

Here $\rho(A,\lambda)$ is the distance between the spectrum $\sigma(A)$ of A and a complex point λ, again.

In order to prove this theorem we will prove a few lemmas.

Lemma 3.4.3 *Let an unbounded quasi-Hermitian operator A admit the triangular representation (2.1) with a spectral resolution $P(t)$. Then there is a sequence $\{A_n\}$ of bounded operators admitting the triangular representation with the same spectral resolution and satisfying the following conditions:*

1. The nilpotent part of A_n coincides with the nilpotent part of A for any $n < \infty$.

2. $\sigma(A_n) \subseteq \sigma(A)$.

3. For all $n < \infty$ the equalities $(A_n)_I = A_I$ and

$$\Im\lambda_k(A_n) = \Im\lambda_k(A) \tag{4.7}$$

hold for each nonreal eigenvalue.

4. Vectors $A_n h$ tend to Ah in the strong topology for any $h \in D(A)$.

Proof: Let $\phi(t)$ be the scalar-valued function defined by (2.2). Denote $h(t) = \Re\phi(t)$ and $\psi(t) = \Im\phi(t)$. Further, set $h_n(t) = h(t)$ if $|h(t)| \leq n$, $h_n(t) = n$ if $h(t) \geq n$ and $h_n(t) = -n$ if $h(t) \leq -n$. Define operators S_n by the formula

$$S_n = \int_{-\infty}^{\infty} (h_n(t) + i\psi(t))dP(t) \tag{4.8}$$

and set $A_n = S_n + V$ $(n = 1, 2, ...)$. Obviously, A_n has the triangular representation with the diagonal part S_n, the nilpotent part V and the spectral resolution $P(t)$.

Since the operator

$$A - A^* = V - V^* + 2i \int_{-\infty}^{\infty} \psi(t)dP(t)$$

is bounded, ψ is bounded on $(-\infty, \infty)$ and, therefore, A_n $(n < \infty)$ is bounded.

Take into account that $\sigma(S_n) \subset \sigma(S)$. By Lemma 3.2.12 this clearly forces $\sigma(A_n) \subset \sigma(A)$.

Further, it is simple to see that

$$A_n - A_n^* = S_n - S_n^* + V - V^* =$$

$$V - V^* + 2i \int_{-\infty}^{\infty} \psi(t)dP(t) = A - A^*.$$

That is, the equality $(A_n)_I = A_I$ is fulfilled, really.

Clearly, $S_n - S_n^* = S - S^*$. This gives $\Im\sigma(S_n) = \Im\sigma(S)$. Now Lemma 3.2.12 yields equality (4.7).

We see at once that by (4.8) vectors $S_n h$ tend to Sh in the strong topology for any $h \in D(S)$. But $D(A) = D(S)$. The proof is complete. $\square$

Lemma 3.4.4 *Let a Volterra operator V satisfy the condition*

$$V_I \equiv \frac{1}{2i}(V - V^*) \text{ is a Hilbert-Schmidt operator.}$$

Then

$$N^2(V) = 2N^2(V_I).$$

Proof: We have

$$Tr V^2 = Tr(V^*)^2 = 0$$

because V^2 is a Volterra operator. Hence

$$N^2(V - V^*) = Tr(V - V^*)^2 = Tr(V^2 + VV^* + V^*V + (V^*)^2)$$
$$= Tr(VV^* + V^*V) = 2Tr(VV^*).$$

We arrive at the result, because

$$N^2(V) = Tr(VV^*).\square$$

Lemma 3.4.5 *Let a quasi-Hermitian operator A satisfy the condition (4.4). Then it admits the triangular representation (due to Theorem 3.4.1), and the relation*

$$N(V) = g_1(A) \tag{4.9}$$

holds where V is the nilpotent part of A.

Proof: First, assume that A is a bounded operator. Let S be the diagonal part of A. From the triangular representation (2.1) it follows that

$$4(A_I)^2 = Tr(A - A^*)^2 = Tr(S + V - S^* - V^*)^2.$$

Using Corollary 3.2.10 and omitting simple calculations, we obtain

$$4(A_I)^2 = Tr(A - A^*)^2 = Tr(S - S^*)^2 + Tr(V - V^*)^2.$$

Set

$$V_I = (V - V^*)/2i \text{ and } S_I = (S - S^*)/2i.$$

That is, we have obtained the equality

$$N^2(A_I) = N^2(V_I) + N^2(S_I).$$

Taking into account Lemma 3.4.4, we arrive at the equality

$$2N^2(A_I) - 2N^2(S_I) = N^2(V).$$

Recall that the nonreal spectrum of a quasi-Hermitian operator consists of isolated eigenvalues. Besides, due to Lemma 3.2.12 $\sigma(A) = \sigma(S)$. Thus,

$$N^2(S_I) = \sum_{k=1}^{\infty} |\Im \lambda_k(A)|^2,$$

and we arrive at the result if A is bounded.

If A is unbounded, then the assertion is due to Lemma 3.4.3 and the just proved result for bounded operators □

Note that we can formulate Lemma 2.11.1 in the following way

Corollary 3.4.6 *Let A be a Hilbert-Schmidt operator. Then the equality*

$$g_1(A) = g(A)$$

is valid.

The assertion of Theorem 3.4.2 follows from Lemma 3.3.1 and Lemma 3.4.5.

Corollary 3.4.7 *Let a linear bounded operator A satisfy the following condition:*

$$A^p - (A^*)^p \text{ is a Hilbert-Schmidt operator}$$

for some integer number p. Then the inequality

$$\|R_\lambda(A)\| \le \|T_{\lambda,p}\| \sum_{k=0}^{\infty} \frac{g_1^k(A^p)}{\rho^{k+1}(A^p, \lambda^p)\sqrt{k!}} \text{ for all regular } \lambda$$

is true.

Here $\rho(A^p, \lambda^p)$ is the distance between the spectrum $\sigma(A^p)$ of A^p and λ^p, and

$$T_{\lambda,p} = \sum_{k=0}^{p-1} A^k \lambda^{p-k-1}.$$

The result follows from Theorem 3.4.2 and the relation

$$R_\lambda(A) = T_{\lambda,p} R_{\lambda^p}(A^p)$$

proved in Section 2.5.

3.4.3 Estimation for resolvents of operators with Neumann-Schatten's imaginary components

In this subsection we will obtain an estimate for the resolvent of generally unbounded quasi-Hermitian operator A, assuming that

$$A_I \in C_{2r} \text{ for some integer } r > 1. \tag{4.10}$$

That is, the inequality

$$N_{2r}(A_I) = \sqrt[2r]{\sum_{j=1}^{\infty} |\lambda_j(A_I)|^{2r}} < \infty$$

holds.

Denote

$$\beta_r = 2(1 + \frac{2r}{exp(2/3)ln2}),$$

i.e. β_r depends on r, only.

Theorem 3.4.8 *Let a quasi-Hermitian operator A acting in H satisfy the condition (4.10). Then the following inequality is valid:*

$$\|R_\lambda(A)\| \le \sum_{m=0}^{r-1} \sum_{k=0}^{\infty} \frac{(\beta_r N_{2r}(A_I))^{kr+m}}{\rho^{rk+m+1}(A,\lambda)\sqrt{k!}} \text{ for all regular } \lambda.$$

In order to prove this theorem we need the following result.

Lemma 3.4.9 *Let A satisfy (4.10), then it admits the triangular represen-tation (due to Theorem 3.4.1), and the nilpotent part V of A satisfies the relation*

$$N_{2r}(V) \leq \beta_r N_{2r}(A_I).$$

Proof: Let V_I , S_I be the imaginary components of V and S, respectively. Here S is the diagonal part of A. According to (2.1) $A_I = V_I + S_I$. First, assume that A is a bounded operator. Due to (1.5) the condition $A_I \in C_{2r}$ entails the inequality

$$N_{2r}(S_I) \leq N_{2r}(A_I).$$

Therefore,

$$N_{2r}(V_I) \leq N_{2r}(A_I) + N_{2r}(S_I) \leq 2N_{2r}(A_I). \qquad (4.11)$$

Due to the well-known Theorem 3.6.2 from the book by Gohberg and Krein (1969) the inequality

$$N_p(V_I) \leq \frac{1}{2}k_p N_p(V + V^*)$$

is valid where the constant k_p depends on p, only. Moreover,

$$k_p \leq \frac{p}{exp(2/3)ln2}.$$

We obtain

$$N_p(V) \leq N_p(V + V^*)/2 + N_p(V_I) \leq N_p(V_I)(1 + k_p).$$

This inequality and (4.11) imply the required result.

Let now A be unbounded. To obtain the stated result in this case it is sufficient apply Lemma 3.4.3 and the just obtained result for bounded operators. $\square$

The assertion of Theorem 3.4.8 follows from Lemma 3.3.1 and Lemma 3.4.9.

3.4.4   Remarks

1. Set for $x, y \geq 0$

$$\theta_r(x,y) \equiv \sum_{m=0}^{r-1}\sum_{k=0}^{\infty}\frac{x^{kr+m}}{y^{kr+m+1}\sqrt{k!}}.$$

Schwarz's inequality gives

$$\theta_r(x,y) \leq \sum_{m=0}^{r-1}\frac{x^m}{y^{m+1}}\Big[\sum_{k=0}^{\infty}(\frac{x}{y})^{2kr}\frac{2^k}{k!}\sum_{j=0}^{\infty}\frac{1}{2^j}\Big]^{1/2}$$

That is,

$$\theta_r(x,y) \leq \sqrt{2}\sum_{m=0}^{r-1}\frac{x^m}{y^{m+1}}exp[(x/y)^{2r}]. \tag{4.12}$$

Therefore, under the condition $A_I \in C_2$ Theorem 3.4.2 implies the inequality

$$\|R_\lambda(A)\| \leq \frac{\sqrt{2}}{\rho(A,\lambda)}exp[\frac{g_1^2(A)}{\rho^2(A,\lambda)}] \text{ for all regular } \lambda. \tag{4.13}$$

Besides, Theorem 3.4.8 gives under the condition $A_I \in C_{2r}$, $r = 2,3,...$ the inequality

$$\|R_\lambda(A)\| \leq \sqrt{2}\sum_{m=0}^{r-1}\frac{\beta_r^m N_{2r}^m(A_I)}{\rho^{m+1}(A,\lambda)}exp[\frac{(\beta_r N_{2r}(A_I))^{2r}}{\rho^{2r}(A,\lambda)}] \tag{4.14}$$

for all regular λ.

2. Let

$$r = 2^n \text{ for } n = 2,3,....$$

Then we can take

$$\beta_r = 2(1 + cot(\pi/4r)).$$

In fact, as shown in Gohberg and Krein (1970, p.124), the following inequality holds

$$N_p(V_I) \leq cot(\pi/2p)N_p(V_R) \text{ for } p = 2^n; n = 1,2,...,$$

where $V_R \equiv (V + V^*)/2$. Further, the inequality

$$N_p(V) \leq N_p(V_R) + N_p(V_I) \leq (1 + cot(\pi/2p))N_p(V_I)$$

and (2.1) yield

$$N_p(V) \leq 2(1 + \cot(\pi/2p))N_p(A_I).$$

Now Lemma 3.3.1 implies the required assertion.

3.5 Norm estimations for resolvents of quasiunitary operators

In this section we will obtain an estimate for the norm of the resolvent of an operator A under the condition

$$AA^* - I \text{ is a nuclear operator.} \qquad (5.1)$$

Besides, our result is extended to the case

$$A^p(A^*)^p - I \text{ is a nuclear operator} \qquad (5.2)$$

for some positive integer number p.

Let (5.1) hold. Set

$$\vartheta(A) = [Tr(A^*A - I) - \sum_{k=1}^{\infty}(|\lambda_k(A)|^2 - 1)]^{1/2}.$$

If A is a unitary operator: $AA^* = A^*A = I$, then $\vartheta(A) = 0$.

Let A have the unitary spectrum, only. This means that $\sigma(A)$ lies on the unit circle. Then

$$\vartheta(A) = [Tr(A^*A - I)]^{1/2}.$$

We will check that if the condition

$$\sum_{k=1}^{\infty}(|\lambda_k(A)|^2 - 1) \geq 0 \qquad (5.3)$$

holds, then $Tr(A^*A - I) \geq 0$ and therefore,

$$\vartheta(A) \leq [Tr(A^*A - I)]^{1/2}. \qquad (5.4)$$

Theorem 3.5.1 *Let an operator A satisfy the condition (5.1) and have a regular point on the unit circle. Then the following inequality*

$$\|R_\lambda(A)\| \le \sum_{k=0}^{\infty} \frac{\vartheta^k(A)}{\sqrt{k!}\rho^{k+1}(A,\lambda)} \text{ for all regular } \lambda \qquad (5.5)$$

is true.

Recall that $\rho(A,\lambda)$ is the distance between $\sigma(A)$ and a complex point λ.

Corollary 3.5.2 *Let an operator A satisfy the conditions (5.1) and (5.3). Then the inequality*

$$\|R_\lambda(A)\| \le \sum_{k=0}^{\infty} \frac{(Tr(A^*A - I))^k}{\sqrt{k!}\rho^{k+1}(A,\lambda)} \text{ for all regular } \lambda$$

is true .

Indeed, this result is due to (5.4) and Theorem 3.5.1.

In order to prove Theorem 3.5.1 we will first prove several lemmas.

Lemma 3.5.3 *Let a bounded operator A admit a triangular representation with a spectral resolution P(t), and let f be a function analytic on a neighborhood of the spectrum of A. Then the operator f(A) admits the triangular representation with the same spectral resolution. Moreover, the diagonal part S_f of $f(A)$ is defined by*

$$S_f = f(S),$$

where S is the diagonal part of A.

Proof: Due to the representation for analytic operator-functions (see Section 2.7) and due to (2.1) we have

$$f(A) = -\frac{1}{2\pi i} \int_C f(\lambda)R_\lambda(A)d\lambda =$$

$$f(A) = -\frac{1}{2\pi i} \int_C f(\lambda)R_\lambda(S)(I + VR_\lambda(S))^{-1}d\lambda.$$

Here the contour C surrounds $\sigma(S)$. Consequently,

$$f(A) = -\frac{1}{2\pi i} \int_C f(\lambda) R_\lambda(S) d\lambda + W = f(S) + W, \qquad (5.6)$$

where

$$W = -\frac{1}{2\pi i} \int_C f(\lambda) R_\lambda(S)[(I + V R_\lambda(S))^{-1} - I] d\lambda.$$

But

$$(I + V R_\lambda(S))^{-1} - I = -V R_\lambda(S)(I + V R_\lambda(S))^{-1}.$$

We thus get

$$W = \frac{1}{2\pi i} \int_C f(\lambda) \psi(\lambda) d\lambda,$$

where

$$\psi(\lambda) = R_\lambda(S) V R_\lambda(S)(I + V R_\lambda(S))^{-1}.$$

Due to Lemma 3.2.4 $\psi(\lambda)$ is a Volterra operator for each $\lambda \in C$ with the same spectral resolution. Since $P(t)$ is a bounded operator, we have by Lemma 3.2.3

$$(P(t_0 + 0) - P(t_0))W(P(t_0 + 0) - P(t_0)) =$$

$$\frac{1}{2\pi i} \int_C f(\lambda)(P(t_0 + 0) - P(t_0))\psi(\lambda)(P(t_0 + 0) - P(t_0)) d\lambda = 0$$

for every gap $P(t_0 + 0) - P(t_0)$ of $P(t)$. Thus, W is a Volterra operator thanks to Lemma 3.2.2. It remains to recall (5.6). $\square$

Lemma 3.5.4 *Let A be a linear operator in H satisfying the condition*

$$A^* A - I \in C_\omega, \qquad (5.7)$$

and let the operator $I - A$ be invertible. Then the operator

$$B = i(I - A)^{-1}(I + A) \qquad (5.8)$$

(Cayley's transformation of A) is bounded and satisfies the condition

$$B - B^* \in C_\omega. \qquad (5.9)$$

Proof: According to (5.7) we can write down

$$A = U(I + K_0) = U + K,$$

where U is a unitary operator and both K and K_0 are compact operators belonging to C_ω. Consequently,

$$B^* = -i(I+U^*+K^*)(I-U^*-K^*)^{-1} = -i(I+U^{-1}+K^*)(I-U^{-1}-K^*)^{-1}.$$

That is,

$$B^* = -i(I + U + K_1)(U - I - K_1)^{-1},$$

where $K_1 = K^*U$. This clearly forces

$$2B_I = (B - B^*)/i = T_1 + T_2,$$

where

$$T_1 = (I + U)[(I - U - K)^{-1} + (U - I - K_1)^{-1}]$$

and

$$T_2 = K(I - U - K)^{-1} + K_1(U - I - K_1)^{-1}.$$

Since $K, K_1 \in C_\omega$, we conclude that $T_2 \in C_\omega$. It remains to prove that $T_1 \in C_\omega$. Let us apply the identity

$$(I-U-K)^{-1}+(U-I-K_1)^{-1} = -(I-U-K)^{-1}(K_1+K)(U-I-K_1)^{-1}.$$

From this we deduce that $T_1 \in C_\omega$ because $K, K_1 \in C_\omega$. This completes the proof. □

Lemma 3.5.5 *Let A be a linear operator in H satisfying the condition (5.7) and let A have a regular point on the unit circle. Then A admits the triangular representation.*

Proof: Without any loss of generality we assume that A has on the unit circle a regular point $\lambda_0 = 1$. In the other case we can consider instead of A the operator $A\lambda_0^{-1}$.

Let us consider in H the operator B defined by (5.8). By the previous lemma it satisfies (4.1) and therefore due to Theorem 3.4.1 has a triangular representation. The transformation inverse to (5.8) must be defined by the formula

$$A = (B - iI)(B + iI)^{-1}.$$

Now Lemma 3.5.3 ensures the result. □

Note that a direct proof of Lemma 3.5.5 can be found in (V. Brodskii, 1970) and (V.Brodskii, Gohberg and Krein, 1969).

Lemma 3.5.6 *Let a linear operator A satisfy (5.1) and have a regular point on the unit circle. Then A has the triangular representation due to Lemma 3.5.5, and the relation*

$$N(V) = \vartheta(A)$$

holds, where V is the nilpotent part of A.

Proof: By Corollary 3.2.10

$$Tr(S^*V) = Tr(V^*S) = 0.$$

Employing the triangular representation (2.1) we obtain

$$Tr(A^*A - I) = Tr[(S + V)^*(S + V) - I)] =$$
$$Tr(S^*S - I) + Tr(V^*V).$$

As it is well known, (see e.g. Gohberg and Krein (1969)), the non-unitary spectrum of a quasiunitary operator consists of eigenvalues which converge to the unit circle. Since S is a normal operator and the spectrums of S and A coincide due to Lemma 3.2.12, then we can write

$$Tr(S^*S - I) = \sum_{k=1}^{\infty}(|\lambda_k(S)|^2 - 1).$$

This equality and the one

$$N^2(V) = Tr(V^*V)$$

imply the result.□

Proof of Theorem 3.5.1: The stated result follows from Lemma 3.3.1 and Lemma 3.5.6.□

Lemma 3.5.7 *Under (5.1) and (5.3) the inequality (5.4) is true.*

Proof: By (1.5)

$$Tr(A * A - I) \geq Tr(S^*S - I) \geq 0.$$

From this the stated result follows. □

Corollary 3.5.8 *Let a linear bounded operator A have under (5.2) a regular point on the unit circle. Then the inequality*

$$\|R_\lambda(A)\| \leq \|T_{\lambda,p}\| \sum_{k=0}^{\infty} \frac{\vartheta^k(A^p)}{\rho^{k+1}(A^p, \lambda^p)\sqrt{k!}} \quad \text{for all regular } \lambda$$

is true.

Here

$$T_{\lambda,p} = \sum_{k=0}^{p-1} A^k \lambda^{p-k-1},$$

and

$$\rho(A^p, \lambda^p) = \inf_{t \in \sigma(A)} |t^p - \lambda^p|$$

is the distance between $\sigma(A^p)$ and the point λ^p, again.

This result is due to the identity

$$R_\lambda(A) = T_{\lambda,p} R_{\lambda^p}(A^p)$$

proved in Section 2.5 and due to Theorem 3.5.1.

Further, Schwarz's inequality gives

$$\sum_{k=0}^{\infty} \frac{\vartheta^k(A)}{\sqrt{k!}\,y^{k+1}} \leq y^{-1} [\sum_{k=0}^{\infty} \frac{\vartheta^{2k}(A))2^k}{y^{2k}k!} \sum_{j=0}^{\infty} (1/2)^j]^{1/2}.$$

Since

$$\sum_{k=0}^{\infty} \frac{2^k \vartheta^{2k}(A)}{y^{2k}k!} = \exp[2\vartheta^2(A)/y^2] \text{ for all } y > 0$$

and

$$\sum_{j=0}^{\infty} (1/2)^j = 2,$$

then Theorem 3.5.1 under (5.1) gives the inequality

$$\|R_\lambda(A)\| \le \frac{\sqrt{2}}{\rho(A,\lambda)} \exp[\frac{\vartheta^2(A)}{\rho^2(A,\lambda)}] \text{ for all regular } \lambda.$$

3.6 Functions of operators admitting the triangular representation

In this section we will establish an estimation for the norm of regular functions of linear operators admitting the triangular representation.

Let A be a linear unbounded operator in a separable Hilbert space with an unbounded spectrum $\sigma(A)$ and $f(z)$ be a scalar-valued function which is analytic on some neighborhood of $\sigma(A)$. Let M be an open set containing $\sigma(A)$ whose boundary C consists of a finite number of Jordan arcs such that f is analytic on $M \cup C$. Let C have positive orientation with respect to M. Then we define the function of the operator A by the equality

$$f(A) = If(\infty) - \frac{1}{2\pi i} \int_C f(\lambda)R_\lambda(A)d\lambda. \tag{6.1}$$

cf. Dunford and Schwartz (1958, p.601).

If A is bounded, then (6.1) is equivalent to the formula

$$f(A) = -\frac{1}{2\pi i} \int_C f(\lambda)R_\lambda(A)d\lambda \tag{6.2}$$

with a closed contour C.

Let P_k be a finite chains of orthogonal projectors:

$$0 = Range(P_0) \subset Range(P_1) \subset \dots \subset Range(P_n) = H$$

We need the following

Lemma 3.6.1 *Let a bounded operator A in H have the representation*

$$A = \sum_{k=1}^n \phi_k \Delta P_k + V$$

where ϕ_k, $(k = 1, ..., n)$ are numbers, and V is a Hilbert-Schmidt operator satisfying the relations

$$P_{k-1}VP_k = VP_k \quad (k = 1, ..., n).$$

Then the inequality

$$\|f(A)\| \le \sum_{k=0}^{n-1} \sup_{\lambda \in co(A)} |f^{(k)}(\lambda)| \frac{N^k(V)}{(k!)^{3/2}}$$

is true for an arbitrary function f holomorphic on a neighborhood of the closed convex hull $co(A)$ of $\sigma(A)$.

Proof: Put

$$S = \sum_{k=1}^{n} \phi_k \Delta P_k.$$

Clearly, the spectrum of S consists of numbers ϕ_k, $(k = 1, ..., n)$.

It is simple to check that $V^n = 0$. That is, V is a nilpotent operator.

Repeating the proof of Lemma 3.2.12 we easily obtain that $\sigma(S) = \sigma(A)$. Consequently, ϕ_k, $(k = 1, ..., n)$ are eigenvalues of A.

Further, let

$$\{e_j(m)\}, \quad m = 1, 2, ...$$

be an orthonormal basis in $\Delta P_j H$. We denote

$$a_{ij}(m) = (Ve_i(m), e_j(m)),$$

and

$$S_m = \sum_{k=1}^{n} \phi_j(., e_j(m))e_j(m).$$

Besides, put

$$V_m = \sum_{1 \le i < j \le n} a_{ij}(m)(., e_j(m))e_i(m).$$

Finally, set

$$A_m = S_m + V_m \quad (m = 1, 2, ...).$$

Clearly, $a_{ij}(m) = 0$ when $i > j$.

The operators A, S, V and $f(A)$ are orthogonal sums of A_m, S_m, V_m and $f(A_m)$ $(m = 1, 2, \ldots)$, respectively. Therefore,

$$\sigma(A_m) \subseteq \sigma(A), \text{ and } N(V_m) \leq N(V).$$

Besides,

$$\max_m \|f(A_m)\| = \|f(A)\|. \tag{6.3}$$

Since A_m is an $n \times n$-matrix, then due to Corollary 1.2.3 we have the inequality

$$\|f(A_m)\| \leq \sum_{k=0}^{n-1} \sup_{\lambda \in co(A_m)} |f^{(k)}(\lambda)| \frac{N^k(V_m)}{(k!)^{3/2}}.$$

This and (6.3) prove the stated result.□

Theorem 3.6.2 *Let a bounded operator A admit the triangular representation. Let also the nilpotent part V of A be a Hilbert-Schmidt operator. Then the inequality*

$$\|f(A)\| \leq \sum_{k=0}^{\infty} \sup_{\lambda \in co(A)} |f^{(k)}(\lambda)| \frac{N^k(V)}{(k!)^{3/2}}$$

is true for an arbitrary function f holomorphic on a neighborhood of $co(A)$.

Proof: Let $P(t)$, $a \leq t \leq b$ be the spectral resolution of A. Define the operators

$$V_n = \sum_{k=0}^{n} P(t_{k-1}) V \Delta P_k$$

and

$$S_n = \sum_{k=1}^{n} \phi(t_k) \Delta P(t_k).$$

$t_k = t_k^{(n)}$, $a = t_0 \leq t_1 \leq \ldots \leq t_n = b$, $\Delta P(t_k) = P(t_k) - P(t_{k-1})$, $k = 1, \ldots, n$.

Besides, put

$$B_n = S_n + V_n.$$

Then the sequence $\{B_n\}$ strongly converges to A due to the triangular representation. According to (6.2) the sequence $\{f(B_n)\}$ strongly converges to $f(A)$. The inequality

$$\|f(A)\| \leq \sup_n \|f(B_n)\| \tag{6.4}$$

is true thanks to the Banach-Steinhaus theorem (see e.g. (Dunford and Schwartz, 1958)). Since the spectral resolution of B_n consists of $n < \infty$ projectors, Lemma 3.6.1 yields the inequality

$$\|f(B_n)\| \leq \sum_{k=0}^{n-1} \sup_{\lambda \in co(B_n)} |f^{(k)}(\lambda)| \frac{N^k(V_n)}{(k!)^{3/2}}. \tag{6.5}$$

Thanks to Lemma 3.2.12 we have $\sigma(B_n) = \sigma(S_n)$. Clearly, $\sigma(S_n) \subseteq \sigma(S)$. Hence,

$$\sigma(B_n) \subseteq \sigma(A). \tag{6.6}$$

Due to Theorem 3.6.3 from (Gohberg and Krein, 1969) $\{N(V_n)\}$ tends to $N(V)$ as n tends to infinity. Thus (6.4), (6.5) and (6.6) imply the result. $\square$

Theorem 3.6.3 *Let an unbounded operator A admit the triangular representation and let its nilpotent part V be a Hilbert-Schmidt operator. Let, additionally, the condition*

$$\inf \Re\sigma(A) > -\infty. \tag{6.7}$$

hold. Then the inequality

$$\|f(A)\| \leq \sum_{k=0}^{\infty} \sup_{\lambda \in co(A)} |f^{(k)}(\lambda)| \frac{N^k(V)}{(k!)^{3/2}}$$

is true for an arbitrary function f holomorphic on a neighborhood of $co(A)$.

Proof: Let $P(t)$, $-\infty \leq t \leq \infty$ be the spectral resolution of A. It is simple to show that under (6.7) $AP(n)$ is the bounded operator for each $n < \infty$. Clearly, that

$$(A - \mu I)^{-1} P(n)(AP(n) - \mu P(n)) = P(n),$$

and therefore,

$$(A - \mu I)^{-1} P(n) = (AP(n) - \mu P(n))^{-1}.$$

According to the representation of the operator-values function (6.2) we can write down $f(AP(n)) = f(A)P(n)$. Consequently, operators $f(AP(n))$ converge to $f(A)$.

Further, since $P(n)$ projects onto an invariant subspace, $\sigma(AP(n)) \subset \sigma(A)$. As proved above, the estimate

$$\|f(AP(n))\| \leq \sum_{k=0}^{\infty} \sup_{\lambda \in co(A)} |f^{(k)}(\lambda)| \frac{N^k(VP(n))}{(k!)^{3/2}}$$

is valid. This proves the stated result.□

3.7 Regular functions of quasi-Hermitian operators

Theorem 3.7.1 *Let a quasi-Hermitian operator A satisfy the conditions (6.7) and*

$$A_I \text{ is a Hilbert-Schmidt operator.} \tag{7.1}$$

Let also f be a holomorphic function on a neighborhood of the closed convex hull co(A) of $\sigma(A)$. Then

$$\|f(A)\| \leq \sum_{k=0}^{\infty} \sup_{\lambda \in co(A)} |f^{(k)}(\lambda)| \frac{g_1^k(A)}{(k!)^{3/2}}. \tag{7.2}$$

Recall that the quantity $g_1(A)$ is defined by the equality (4.2).

Proof: The stated result immediately follows from Theorem 3.6.3 and Lemma 3.4.5. □

Theorem 3.7.1 is precise: the inequality (7.2) becomes the equality

$$\|f(A)\| = \sup_{\mu \in \sigma(A)} |f(\mu)| \tag{7.3}$$

if A is a normal operator and

$$\sup_{\lambda \in co(A)} |f(\lambda)| = \sup_{\lambda \in \sigma(A)} |f(\lambda)|, \tag{7.4}$$

because $g_1(A) = 0$ for a normal operator A.

Example 3.7.2 *Let a bounded operator A satisfy the condition (7.1). Then the inequality*

$$\|A^m\| \le \sum_{k=0}^{m} \frac{m! d^{m-k}(A) g_1^k(A)}{(m-k)!(k!)^{3/2}}$$

holds for any integer m. Recall that $d(A)$ is the spectral radius of A.

Example 3.7.3 *Let a generally unbounded operator A satisfy the conditions (7.1) and*

$$\alpha(A) = \sup \Re\sigma(A) < \infty. \tag{7.5}$$

Then the inequality

$$\|e^{At}\| \le e^{\alpha(A)t} \sum_{k=0}^{\infty} \frac{t^k g_1^k(A)}{(k!)^{3/2}} \text{ for all } t \ge 0$$

holds.

Indeed (7.5) coincides with (6.7) if A is replayed by $-A$.

3.8   Functions of quasiunitary operators

Theorem 3.8.1 *Let a linear operator A satisfy the condition*

$$AA^* - I \text{ is a nuclear operator,} \tag{8.1}$$

and let A have a regular point on the unit circle. If f is a holomorphic function on a neighborhood of closed convex hull $co(A)$ of $\sigma(A)$, then the estimate

$$\|f(A)\| \le \sum_{k=0}^{\infty} \sup_{\lambda \in co(A)} |f^{(k)}(\lambda)| \frac{\vartheta^k(A)}{(k!)^{3/2}} \tag{8.2}$$

is valid.

Recall that the quantity $\vartheta(A)$ is defined in Section 3.5.

Proof: The result immediately follows from Lemma 3.5.6 and Theorem 3.6.2. □

Theorem 2.8.1 is precise: the inequality (8.2) becomes the equality (7.3) if A is a unitary operator and (7.4) holds because $\vartheta(A) = 0$ in this case.

Example 3.8.2 *Let an operator A satisfy the condition (8.1) and have a regular point on the unit circle. Then the inequality*

$$\|A^m\| \le \sum_{k=0}^{m} \frac{m! d^{m-k}(A) \vartheta^k(A)}{(m-k)!(k!)^{3/2}}$$

holds for any integer m. Recall that $d(A)$ is the spectral radius of A.

3.9 Multiplicative representations of resolvents of nonselfadjoint operators

3.9.1 General spectral resolutions

Recall that the notion of the multiplicative integral is presented in Subsection 2.8.2.

Theorem 3.9.1 *Let a linear operator A acting in a separable Hilbert possess a spectral resolution $P(t)$ $(a \le t \le b)$ and the triangular representation (2.1). Then for any regular λ the equality*

$$R_\lambda(A) = \int_a^b \frac{dP(s)}{\phi(s) - \lambda} \int_{[a,s]}^{\longrightarrow} (I - \frac{1}{\phi(t) - \lambda} V dP(t)) \tag{9.1}$$

is true, where V is the nilpotent part of A.

This means that $R_\lambda(A)$ is the limit in the strong topology of the sequence of the operators

$$\sum_{k=1}^{n} \frac{\Delta P(t_k)}{\phi(t_k) - \lambda} \prod_{k+1 \le j \le n}^{\longrightarrow} (I + \frac{1}{\phi(t_j) - \lambda} V \Delta P(t_j))$$

for the corresponding finite partitioning $\{t_k\}$ of $[a, b]$.
Proof: Exploiting (2.1) we have

$$R_\lambda(A) = R_\lambda(S)(I + V R_\lambda(S))^{-1}. \tag{9.2}$$

By Lemma 3.2.11 $V R_\lambda(S)$ is a Volterra operator. We invoke Lemma 2.8.7. It asserts that

$$(I + V R_\lambda(S))^{-1} = \int_{[a,b]}^{\longrightarrow} (I - V R_\lambda(S) dP(t)).$$

But according to (2.2)

$$R_\lambda(S) = \int_a^b \frac{dP(s)}{\phi(s) - \lambda}. \tag{9.3}$$

Thus,

$$R_\lambda(S)dP(t) = \frac{1}{\phi(t) - \lambda}dP(t).$$

From (9.2) it follows

$$R_\lambda(A) = \int_a^b \frac{dP(s)}{\phi(s) - \lambda} \int_{[a,b]}^{\longrightarrow} (I - \frac{1}{\phi(t) - \lambda}VdP(t)) \tag{9.4}$$

for all regular λ. But

$$dP(s)VdP(t) = 0 \text{ for } t \leq s. \tag{9.5}$$

Thus the stated result follows from (9.4) .□

Let us suppose that A is a normal operator. Then $V = 0$ and Theorem 3.9.1 yields the classical representation (9.3) for the resolvent of normal operators (Dunford and Schwartz, 1963).

Corollary 3.9.2 *Under the hypothesis of Theorem 3.9.1 for all regular λ the representation*

$$R_\lambda(A) = \int_a^b \frac{dP(s)}{\phi(s) - \lambda} \int_{[s,b]}^{\longrightarrow} (I - \frac{2i}{\phi(t) - \lambda}[P(t)A_I - \Im\phi(t)]dP(t)) \tag{9.6}$$

is valid.

Indeed, due to (2.1)

$$A_I dP(t) = V_I dP(t) + S_I dP(t).$$

But

$$P(t)VdP(t) = VdP(t), \tag{9.7}$$

and $dP(t)VP(t) = dP(t)VdP(t) = 0$. Then

$$P(t)V^*dP(t) = 0. \tag{9.8}$$

Thus,

$$VdP(t) = 2iP(t)V_I dP(t).$$

It remains to take into account that

$$S_I dP(t) = \Im\phi(t)dP(t)$$

and to apply Theorem 3.9.1.

Further, let A have a purely real spectrum. Then by (9.6) the representation

$$R_\lambda(A) = \int_a^b \frac{dP(s)}{\phi(s) - \lambda} \int_{[s,b]}^{\longrightarrow} (I - \frac{2i}{\phi(t) - \lambda} P(t) A_I dP(t))$$

for all regular λ is valid.

Note that (9.7) and (9.8) give us the equality

$$VdP(t) = 2P(t)V_R dP(t).$$

Due to (2.1)

$$A_R = V_R + S_R.$$

Here A_R, V_R and S_R are the real components of A, V and S, respectively. Now by Theorem 3.9.1 we easily obtain the following:

Corollary 3.9.3 *Under the hypothesis of Theorem 3.9.1, the representation*

$$R_\lambda(A) = \int_a^b \frac{dP(s)}{\phi(s) - \lambda} \int_{[s,b]}^{\longrightarrow} (I - \frac{2}{\phi(t) - \lambda} [P(t)A_R - \Re\phi(t)]dP(t))$$

for all regular λ is valid.

3.9.2 Operators with discrete spectral resolutions

Let a bounded linear operator A possess a purely discrete spectral resolution. This means that there is a sequence of orthogonal projectors $P_0, P_1, P_2, \ldots$ with the following properties:

1. $0 = Range(P_0) \subset Range(P_1) \subset Range(P_2) \subset \ldots$.

2. $P_\infty = \lim_{k\to\infty} P_k = I$.

3. $P_k A P_k = A P_k$, $k = 1, 2, \dots$.

4. Operators $\Delta P_k = P_k - P_{k-1}$, $k = 1, 2, \dots$ project onto one-dimensional subspaces.

Let us suppose that A has a triangular representation with a discrete spectral resolution P_k. This means that $A = S + V$, where

$$S = \sum_{k=1}^{\infty} \lambda_k(S) \Delta P_k$$

and V is a Volterra operator satisfying

$$P_{k-1} V P_k = V P_k, \quad k = 1, 2, \dots.$$

Note that due to Lemma 3.2.12 $\lambda_k(S) = \lambda_k(A)$. Besides,

$$\Delta P_k V \Delta P_k = 0, \quad k = 1, 2, \dots.$$

Hence,

$$\Delta P_k A \Delta P_k = \lambda_k(A), \quad k = 1, 2, \dots.$$

Let e_k be a normed eigenvector of P_k. Then we easily obtain

$$(A e_k, e_j) = 0 \text{ if } j > k \text{ and } (A e_k, e_k) = \lambda_k(A).$$

That is, $\{e_k\}$ is the Schur basis of the operator A.

Theorem 3.9.4 *Let a linear operator A admit the triangular representation with a discrete spectral resolution $\{P_k\}$ ($k = 1, 2, \dots$. Then for all regular λ the equality*

$$R_\lambda(A) = \sum_{k=1}^{\infty} \frac{\Delta P_k}{\lambda_k(A) - \lambda} \overrightarrow{\prod_{k+1 \le j \le \infty}} \left(I - \frac{V \Delta P_j}{\lambda_j(A) - \lambda} \right)$$

is valid.

The operator in the right part is understood as the strong limit of the operators

$$\sum_{k=1}^{n} \frac{\Delta P_k}{\lambda_k(A) - \lambda} \overrightarrow{\prod_{k+1 \le j \le n}} \left(I - \frac{V \Delta P_j}{\lambda_j(A) - \lambda} \right)$$

as n tends to infinity.

The proof of this theorem is based on Theorem 3.9.1. The details are left to the reader.

One can rewrite Corollaries 3.9.2 and 3.9.3 in the case of a purely discrete spectral resolution in the following way: for any regular λ

$$R_\lambda(A) = \sum_{k=1}^\infty \frac{\Delta P_k}{\lambda_k(A) - \lambda} \overrightarrow{\prod_{k+1 \leq j \leq \infty}} (I - \frac{2i[P_j A_I - \Im\lambda_j(A)]\Delta P_j}{\lambda_j(A) - \lambda})$$

and

$$R_\lambda(A) = \sum_{k=1}^\infty \frac{\Delta P_k}{\lambda_k(A) - \lambda} \overrightarrow{\prod_{k+1 \leq j \leq \infty}} (I - \frac{2[P_j A_R - \Re\lambda_j(A)\Delta P_j]}{\lambda_j(A) - \lambda}).$$

3.10   Notes

The notion of the triangular representation of nonselfadjoint operators was developed by many authors (see Livsic (1966), Gohberg and Krein (1970), Brodskii (1971), L. de Branges (1965) etc).

Definition 3.2.2 is taken from Gil'(1993c).

The contents of Section 3.3 are adapted from Gil' (1983a) and Gil' (1993c).

The material in **Section 3.4** is taken from Gil' (1983a), Gil' (1992) and Gil' (1993c). As mentioned above, Theorem 3.4.1 is based on the fundamental results by L.de Branges (1965). Lemma 3.4.4 is due to L. Sahnovich (see Gohberg and Krein (1970)).

Section 3.5 contains the main result from Gil' (1992). The material in **Section 3.6 and Section 3.7** is taken from Gil' (1992). Theorem 3.8.1 is new.

The contents of Section 3.9 are based on Gil' (1973b).

Chapter 4

Perturbations of Finite Dimensional and Compact Operators

This chapter is devoted to spectrum perturbations of finite-dimensional and compact operators. Our main tools are the estimations for the norms of the resolvents obtained above. By these estimations, bounds for the spectra of perturbed operators are obtained.

In Sections 4.1 and 4.2 the definitions and preliminary results are presented.

Section 4.3 deals with finite-dimensional operators. A bound for the spectra of a perturbed operator is established. In addition, the behavior of the multiplicities is investigated.

By the multiplicative structure of the resolvent of a finite dimensional operator in Sections 4.4 and 4.5, bounds for eigenvalues and for the determinant of an arbitrary finite matrix are derived.

Section 4.6 is devoted to approximations of a compact operator by finite dimensional operators. A bound for eigenvalues of a compact operator is

125

obtained.

In Section 4.7 spectrum perturbations of Hilbert-Schmidt operators are examined.

In Section 4.8 compact operators with a finite trace are investigated. A bound for the determinant is derived.

Section 4.9 deals with operators whose powers are Hilbert-Schmidt operators. A bound for the spectra of a perturbed operator is established. Besides, the behavior of the multiplicities is investigated.

In Section 4.10 operators belonging to the Neumann-Schatten ideals are examined.

By the multiplicative structure of the resolvent of a compact operator with a finite trace, in Section 4.11 bounds for eigenvalues are derived.

4.1   Preliminaries

We recall some well-known definitions from the matrix perturbation theory (see (Stewart and Sun, 1990)).

Let a matrix A of the order n have eigenvalues $\lambda_1(A), ..., \lambda_n(A)$ and a matrix B of the order n have eigenvalues $\lambda_1(B), ..., \lambda_n(B)$. Then the spectral variation of B with respect to A is

$$sv_A(B) \equiv \max_i \min_j |\lambda_i(B) - \lambda_j(A)|.$$

We extend this definition.

Let now A and B be linear operators in a Banach space X and $\sigma(A)$ be the spectrum of A.

Definition 4.1.1 *We will say that the quantity*

$$sv_A(B) \equiv \sup_{\mu(B)\in\sigma(B)} \inf_{\lambda(A)\in\sigma(A)} |\mu(B) - \lambda(A)| \qquad (1.1)$$

is the spectral variation of a linear operator B with respect to a linear operator A.

Definition 4.1.2 *The Hausdorff distance between the spectra of A and B is*

$$hd(A, B) \equiv \max\{sv_A(B), sv_B(A)\}. \tag{1.2}$$

First, we will prove a few technical lemmata concerning spectral perturbations.

Lemma 4.1.3 *Let A, T and L be linear operators in a Banach space and let L have a bounded inverse operator. If*

$$q(L) \equiv \| (A - T)L^{-1} \| < \infty,$$

then each point $\mu \in \sigma(T)$ satisfies the inequality

$$q(L) \| LR_\mu(A) \| \geq 1. \tag{1.3}$$

Proof: Suppose that for some $\mu \in \sigma(T)$, the inequality

$$\delta(L) \equiv q(L) \| LR_\mu(T) \| < 1 \tag{1.4}$$

holds. We can write

$$R_\mu(A) - R_\mu(T) = R_\mu(T)(T - A)R_\mu(A) =$$

$$R_\mu(T)(A - T)LL^{-1}R_\mu(A).$$

From this it follows that

$$\| R_\mu(A) - R_\mu(T) \| \leq \| R_\mu(T) \| \delta(L).$$

That is,

$$\| R_\mu(T) \| \leq \| R_\mu(A) \| + \| R_\mu(T) \| \delta(L).$$

The condition (1.4) implies:

$$\| R_\mu(T) \| \leq \| R_\mu(A) \| (1 - \delta(L))^{-1}. \tag{1.5}$$

Thus (1.4) entails that μ is a regular point of T. This contradiction proves the result.□

Lemma 4.1.4 *Let A and B be linear operators in a Banach space, and let*

$$q \equiv \| A - B \| < \infty.$$

Assume that

$$\| R_\lambda(A) \| \leq \phi(\rho^{-1}(A, \lambda)) \text{ for all regular } \lambda \text{ of } A,$$

where $\phi(x)$ is a monotonically increasing non-negative function of a non-negative variable x, such that

$$\phi(0) = 0 \quad and \quad \phi(\infty) = \infty.$$

Then the inequality

$$sv_A(T) \leq z(\phi, q) \tag{1.6}$$

is true, where $z(\phi, q)$ is the extreme right-hand (positive) root of the equation

$$1 = q\phi(1/z). \tag{1.7}$$

Proof: Lemma 4.1.3 gives the inequality

$$1 \leq q \| R_\lambda(A) \| \text{ for all } \lambda \in \sigma(B).$$

From this it follows that

$$1 \leq q\phi(\rho^{-1}(A, \lambda)) \text{ for all } \lambda \in \sigma(B). \tag{1.8}$$

Compare this inequality with (1.7). Since $\phi(x)$ monotonically increases, $z(\phi, q)$ is a unique positive root of (1.7), and $\rho(A, \lambda) \leq z(\phi, q)$. That is, the inequality (1.6) is valid. □

4.2 Perturbations of multiplicities

In this section we will derive a perturbation result for the multiplicities of eigenvalues. Recall that an eigenvalue $\lambda(A)$ has the algebraic multiplicity ν if the equation $Ah = \lambda(A)h$ has ν different non-zero solutions.

In this section H denotes a separable Hilbert space, and $\Omega(c, a)$ denotes the disc on the complex plane with the center $c \in \mathbf{C}$ and the radius a, i.e.

$$\Omega(c, a) \equiv \{z \in \mathbf{C} : |z - c| \leq a\}.$$

Lemma 4.2.1 *Let A and B be linear operators in H. Suppose that*

$$q = \|A - B\| < \infty \qquad (2.1)$$

and

$$\| R_\lambda(A) \| \le \phi(\frac{1}{\rho(A,\lambda)}) \text{ for all regular } \lambda, \qquad (2.2)$$

where $\phi(x)$ is a monotonically increasing non-negative function of a non-negative variable x. Let, in addition, A have an eigenvalue $\lambda(A)$ of the algebraic multiplicity ν and the distance from $\lambda(A)$ to the rest of $\sigma(A)$ is equal to d, i.e.

$$distance\{\lambda(A), \sigma(A)/\lambda(A)\} = d. \qquad (2.3)$$

If there is a positive number $a < d$ such that the relation

$$0 \le \frac{qa\phi^2(1/a)}{1 - q\phi(1/a)} < 1 \qquad (2.4)$$

holds, then the operator B has in $\Omega(\lambda(A), a)$ eigenvalues whose total algebraic multiplicity is equal to ν.

Proof: Let $\partial\Omega(\lambda(A), a) \equiv \partial\Omega$ be the boundary of $\Omega(\lambda(A), a)$, i.e.

$$\partial\Omega \equiv \{z \in \mathbb{C} : |z - \lambda(A)| = a\}.$$

According to (2.2):

$$\| R_\lambda(A) \| \le \phi(1/a) \text{ for each } \lambda \in \partial\Omega.$$

The relationship (2.4) implies

$$q\phi(1/a) < 1. \qquad (2.5)$$

Thus,

$$q \| R_\lambda(A) \| \le q\phi(1/a) < 1 \text{ for each } \lambda \in \partial\Omega. \qquad (2.6)$$

From (1.5) it follows that

$$\|R_\lambda(B)\| \le \frac{\|R_\lambda(A)\|}{1 - q\phi(1/a)} \text{ for each } \lambda \in \partial\Omega. \qquad (2.7)$$

Hence, B has no points of the spectrum which lie on $\partial\Omega$. Set

$$P(A) = -\frac{1}{2\pi i}\int_{\partial\Omega}R_\lambda(A)d\lambda$$

and

$$P(B) = -\frac{1}{2\pi i}\int_{\partial\Omega}R_\lambda(B)d\lambda.$$

That is, both $P(A)$ and $P(B)$ are the projectors onto the eigenspaces corresponding to points of the spectra of A and B which lie in $\Omega(\lambda(A), a)$ (Dunford and Schwartz, 1958).

We have

$$\|P(A) - P(B)\| \le \frac{1}{2\pi}\int_{|\lambda-\lambda(A)|=a}\|R_\lambda(A) - R_\lambda(B)\|d|\lambda|.$$

But

$$\|R_\lambda(A) - R_\lambda(B)\| = \|R_\lambda(A)(A - B)R_\lambda(B)\| \le$$

$$q\|R_\lambda(A)\|\,\|R_\lambda(B)\|.$$

According to (2.7) and (2.6)

$$\|P(A) - P(B)\| \le \frac{qa\phi^2(1/a)}{1 - q\phi(1/a)}.$$

The relationship (2.4) implies

$$\|P(A) - P(B)\| < 1.$$

Thanks to Sz.-Nagy's lemma (Gohberg and Krein, 1969, Lemma 1.3.1), the spaces $P(A)H$ and $P(B)H$ have the same dimensions. But the dimension of $P(A)H$ is equal to ν. We arrive at the result. $\square$

Assume that $\sigma(A)$ consists of eigenvalues $\lambda_k(A)$, only, and the minimal distance between different eigenvalues is equal to d, i.e.

$$|\lambda_k(A) - \lambda_j(A)| \ge d > 0 \; j, k = 1, 2, ..., n_1. \tag{2.8}$$

for all the different eigenvalues. Here $n_1 \le \infty$ is *the number of the different eigenvalues of A* .

Corollary 4.2.2 *Let the spectrum of a linear operator A in H consist of eigenvalues $\lambda_k(A)$, only. Let the corresponding multiplicities be equal to ν_k. Let also (2.1), (2.2), and (2.8) hold. If, in addition, there is a positive number $a < d$ such that (2.4) holds, then $\sigma(B)$ consists of eigenvalues, only. All of them lie in the set*

$$\cup_{k=1}^{n_1} \Omega(\lambda_k(A), a),$$

where

$$\Omega(\lambda_k(A), a) \equiv \{z \in \mathbf{C} : |z - \lambda_k(A)| \le a\}.$$

Moreover, the total algebraic multiplicity of eigenvalues which lie in each $\Omega(\lambda_k(A), a)$ is equal to ν_k.

Lemma 4.2.3 *Let A and B be linear operators in H satisfying (2.1), (2.2). Suppose A has an eigenvalue $\lambda(A)$ of the algebraic multiplicity ν with the property (2.3). Let, in addition, the following inequality*

$$d > z_0(q) \tag{2.9}$$

hold, where $z_0(q)$ is the extreme right-hand (unique positive and simple) root of the scalar equation

$$q[z\phi^2(1/z) + \phi(1/z)] = 1. \tag{2.10}$$

Then operator B has in the disc

$$\Omega(\lambda(A), z_0(q)) \equiv \{z \in \mathbf{C} : |z - \lambda(A)| \le z_0(q)\}$$

the eigenvalues whose total algebraic multiplicity is equal to ν.

Proof: Rewrite (2.10) in the form

$$\frac{qz\phi^2(1/z)}{1 - q\phi(1/z)} = 1. \tag{2.11}$$

Take for sufficiently small $\epsilon > 0$, the number $a = z_0(q) + \epsilon$. Then due to (2.11) the condition (2.4) is satisfied. Thus the result follows from Lemma 4.2.1 since ϵ is arbitrary. $\square$

Corollary 4.2.4 *Let the spectrum of a linear operator A in H consist of eigenvalues $\lambda_k(A)$, only. Let the corresponding multiplicities be equal to ν_k. Let also (2.1), (2.2), (2.8), and (2.9) hold. Then $\sigma(B)$ consists of eigenvalues, only. All of them lie in the set*

$$\cup_{k=1}^{n_1}\Omega(\lambda_k(A), z_0(q)),$$

where

$$\Omega(\lambda_k(A), z_0(q)) \equiv \{z \in \mathbf{C} : |z - \lambda_k(A)| \le z_0(q)\}.$$

Moreover, the total algebraic multiplicity of eigenvalues belonging to each $\Omega(\lambda_k(A), a)$ is equal to ν_k.

4.3 Spectrum perturbations of finite dimensional operators

4.3.1 Perturbations of eigenvalues

We recall that the quantities $g(A)$ and $\gamma_{n,k}$ are defined in Section 1.1.

Theorem 4.3.1 *Let A and B be $n \times n$ matrices. Then the following bound is true:*

$$sv_A(B) \equiv \max_i \min_j |\lambda_i(B) - \lambda_j(A)| \le z(q, A), \qquad (3.1)$$

where $q = \|A - B\|$ and $z(q, A)$ is the extreme right-hand (unique positive and simple) root of the algebraic equation

$$z^n = q\sum_{j=0}^{n-1}\gamma_{n,j} z^{n-j-1} g^j(A). \qquad (3.2)$$

Proof: Theorem 1.2.2 gives

$$\|R_\lambda(A)\| \le \sum_{k=0}^{n-1}\frac{g^k(A)\gamma_{n,k}}{\rho^{k+1}(A,\lambda)} \text{ for all regular } \lambda. \qquad (3.3)$$

Due to Lemma 4.1.4

$$sv_A(B) \le z_1$$

where z_1 is the extreme right-hand (unique positive and simple) root of the equation

$$1 = \sum_{j=0}^{n-1} \frac{\gamma_{n,j} g^j(A)}{z^{j+1}}.$$

Multiplying this equation by z^n, we obtain the equation (3.2). That is, z_1 coincides with $z(q, A)$. The proof is complete. $\square$

Put

$$w(n) = \sum_{j=0}^{n-1} \gamma_{n,j}.$$

We will prove that under the condition

$$qw(n) \geq g(A)$$

the inequality

$$z(q, A) \leq qw(n), \tag{3.4}$$

holds, and under the condition

$$qw(n) \leq g(A)$$

the inequality

$$z(q, A) \leq g^{1-1/n}(A)[qw(n)]^{1/n} \tag{3.5}$$

is valid.

To prove the estimates (3.4) and (3.5) let us consider the algebraic equation of the shape

$$z^n = P(z), \tag{3.6}$$

where $P(z)$ is the polynomial

$$P(z) = \sum_{j=0}^{n-1} c_j z^{n-j-1}$$

with non-negative coefficients c_j.

Lemma 4.3.2 *The extreme right-hand root z_0 of equation (3.6) is non-negative and the following estimates*

$$z_0 \leq [P(1)]^{1/n} \quad if \, P(1) \leq 1, \tag{3.7}$$

and

$$z_0 \leq P(1) \quad if \; P(1) \geq 1 \tag{3.8}$$

are valid.

Proof: Since all the coefficients of $P(z)$ are non-negative, it does not decrease as $z > 0$ increases. From this it easily follows that if $P(1) \leq 1$, then $z_0 \leq 1$ and (3.6) implies (3.7). Let now $P(1) \geq 1$, then due to (3.6) $z_0 \geq 1$, because $P(z)$ does not decrease. It is clear that

$$P(z_0) \leq z_0^{n-1} P(1)$$

in this case. Substituting this inequality in (3.6), we get the inequality (3.8). $\square$

Let $z(q, A)$ be the extreme right-hand root of (3.2), again. Setting $z = g(A)y$ in (3.2) and applying Lemma 4.3.2 we obtain the estimations (3.4) and (3.5).

Theorem 4.3.1 and the estimations (3.4) and (3.5) ensure the following result.

Corollary 4.3.3 *Let the condition*

$$qw(n) \geq g(A)$$

hold, then

$$sv_A(B) \leq qw(n).$$

If the condition

$$qw(n) \leq g(A)$$

holds, then

$$sv_A(B) \leq g^{1-1/n}(A)[qw(n)]^{1/n}.$$

The following result is due to Theorem 4.3.1 and the definition of the Hausdorff distance between the spectra.

Theorem 4.3.4 *Let A and B be $n \times n$ matrices, then the inequality*

$$hd(A, B) \equiv \max\{sv_A(B), sv_B(A)\} \leq \max\{z(q, A), z(q, B)\}$$

is valid, where $z(q, A)$ and $z(q, B)$ are the extreme right-hand (unique positive and simple) roots of (3.2) and the algebraic equation

$$z^n = q \sum_{j=0}^{n-1} \gamma_{n,j} z^{n-j-1} g^j(B),$$

respectively.

Set

$$m(A, B) = max\{g(A), g(B)\}.$$

By Lemma 4.3.2 and Theorem 4.3.4 we easily obtain

Corollary 4.3.5 *Let the condition*

$$qw(n) \geq m(A, B)$$

hold, then

$$h(A, B) \leq qw(n).$$

If the condition

$$qw(n) \leq m(A, B)$$

holds, then

$$hd(A, B) \leq m^{1-1/n}(A, B) \, [qw(n)]^{1/n}.$$

4.3.2   Perturbations of multiplicities of matrices

Set

$$G_n(x) \equiv \sum_{j=0}^{n-1} \gamma_{n,j} x^j g^j(A)$$

for a positive scalar variable x .

Theorem 4.3.6 *Let A and B be n × n-matrices, and let A have an eigenvalue $\lambda(A)$ of the algebraic multiplicity ν such that the distance from $\lambda(A)$ to the rest of $\sigma(A)$ is equal to d, i.e.(2.3) holds.*

If, in addition, there is a positive number $a < d$ such that

$$0 \le \frac{aqG_n^2(1/a)}{1 - qG_n(1/a)} < 1, \tag{3.9}$$

then the matrix B has in $\Omega(\lambda(A), a)$ eigenvalues whose total algebraic multiplicity is equal to ν.

Proof: In fact, this result immediately follows from Lemma 4.2.1 and the estimate (3.3).□

Denote by n_1 *the number of the different eigenvalues of A*, i.e.

$$\sum_{k=1}^{n_1} \nu_k = n,$$

where ν_k are the multiplicity of $\lambda_k(A)$.

Corollary 4.3.7 *Let A and B be n × n-matrices, and let A have the eigenvalues $\lambda_k(A)$ of the multiplicities ν_k, $(k = 1, 2, ..., n_1)$ and (2.8) hold. Suppose, in addition, that there is a positive number $a < d$ such that (3.9) holds. Then all the eigenvalues $\lambda_j(B)$ of B lie in the set*

$$\cup_{k=1}^{n_1} \Omega(\lambda_k(A), a),$$

where

$$\Omega(\lambda_k(A), a) \equiv \{z \in \mathbf{C} : |z - \lambda_k(A)| \le a\}.$$

Moreover, the total algebraic multiplicity of the eigenvalues of B that lie in each $\Omega(\lambda_k(A), a)$ is equal to ν_k.

Theorem 4.3.8 *Let A and B be n × n matrices, and let A have an eigenvalue $\lambda(A)$ of the algebraic multiplicity ν. Let the distance from $\lambda(A)$ to the rest of $\sigma(A)$ be equal to d. Let, additionally,*

$$d > z_1(q, A) \tag{3.10}$$

where $z_1(q, A)$ is the extreme right-hand (unique positive and simple) root of the algebraic equation

$$z^{2n-1} = qz^{2n-1}[zG_n^2(1/z) + G_n(1/z)]. \qquad (3.11)$$

Then B has in the ball

$$\Omega(\lambda(A), z_1(q, A)) \equiv \{z \in \mathbf{C} : |z - \lambda(A)| \leq z_1(q, A)\}$$

the eigenvalues whose total algebraic multiplicity is equal to ν.

Proof: Dividing (3.11) by z^{2n-1}, we obtain the equation

$$q[zG_n^2(1/z) + G_n(1/z)] = 1.$$

Now the result follows from Lemma 4.2.3 and the estimate for the resolvent (3.3).□

Corollary 4.3.9 *Let A and B be $n \times n$-matrices. Let (2.8) and (3.10) hold. Then all the eigenvalues of B lie in the set*

$$\cup_{k=1}^{n_1} \Omega(\lambda_k(A), z_1(q, A))$$

where

$$\Omega(\lambda_k(A), z_1(q, A)) \equiv \{z \in \mathbf{C} : |z - \lambda_k(A)| \leq z_1(q, A)\}.$$

4.4 Bounds for eigenvalues of matrices

4.4.1 Statement of the result

Let A be an $n \times n$-matrix with the entries $a_{jk}; j, k = 1, ..., n$. Denote

$$q_T = \max_{j=2,...,n} \sum_{k=1}^{j-1} |a_{jk}|$$

and

$$m(k, A) = \max_{j=1,...,k-1} |a_{jk}| \quad \text{for} \quad k = 2, ..., n.$$

Theorem 4.4.1 *All the eigenvalues of any $n \times n$-matrix $A = (a_{jk})$ lie in the set*

$$\cup_{j=2}^{n} \Omega(a_{kk}, z(A)),$$

where

$$\Omega(a_{kk}, z(A)) = \{\lambda \in \mathbf{C} : |\lambda - a_{kk}| \le z(A)\} \ ; k = 1, ..., n,$$

and $z(A)$ is the extreme right-hand (unique positive and simple) root of the algebraic equation

$$z^n = q_T \prod_{k=2}^{n} (z + m(k, A)). \tag{4.1}$$

The proof is given in the next subsection.

Corollary 4.4.2 *Let*

$$|a_{kk}| > z(A), \ k = 1, 2, ..., n.$$

Then A is invertible.

Indeed, Theorem 4.4.1 implies that 0 cannot belong to the spectrum of A under the imposed conditions.

Set

$$W(n) = \prod_{k=2}^{n} (1 + m(k, A)).$$

From Lemma 4.3.2 about a root of a polynomial it follows that

$$z(A) \le q_T W(n) \text{ if } q_T W(n) > 1,$$

and

$$z(A) \le \sqrt[n]{q_T W(n)} \text{ if } q_T W(n) \le 1.$$

These estimates and Theorem 4.4.1 entail the following

Corollary 4.4.3 . *Let $A = (a_{jk})$ be an $n \times n$-matrix. If $q_T W(n) \le 1$, then all the eigenvalues of A lie in the union of the discs*

$$\{\lambda \in \mathbf{C} : |\lambda - a_{kk}| \le \sqrt[n]{q_T W(n)}\} \ ; k = 1, ..., n.$$

If $q_T W(n) > 1$, then all eigenvalues of A lie in the union of the discs

$$\{\lambda \in C : |\lambda - a_{kk}| \le q_T W(n) \} ; k = 1, ..., n.$$

4.4.2 Proof of Theorem 4.4.1

Let the norm $\|A\|_0$ of the matrix A be defined as the maximum of the sums of moduli of the column elements, i.e.

$$\|A\|_0 = \max_{j=1,\ldots,n} \sum_{k=1}^{n} |a_{jk}|.$$

Now consider an upper-triangular matrix

$$T = (t_{jk}) \text{ with } t_{jk} = 0 \text{ if } j > k. \tag{4.2}$$

Lemma 4.4.4 *Let T be an upper-triangular matrix with the elements (4.2). Then the following estimate is true:*

$$\|R_\lambda(T)\|_0 \leq \rho^{-1}(T,\lambda) \prod_{k=2}^{n} (1 + \frac{m(k,T)}{|t_{kk} - \lambda|}),$$

where

$$\rho(T,\lambda) \doteq \min_{k=1,\ldots,n} |t_{kk} - \lambda|$$

and

$$m(k,T) \doteq \max_{j=2,\ldots,k-1} |t_{jk}|.$$

Proof: Let $V(T) = (v_{jk})$ be the nilpotent part of T, i.e.

$$v_{kk} = 0, \ k = 1, \ldots, n$$

and

$$v_{jk} = t_{jk} \text{ for all } j < k.$$

Further, let $P_k, k = 0, \ldots, n$, $P_0 = 0, P_n = I$ be the increasing chain of the projectors onto the invariant subspaces of T. Then

$$\|V(T)\Delta P_k\|_0 = m(k,T).$$

Since t_{kk} are eigenvalues of the upper-triangular matrix T, then the diagonal part $D(T)$ of T is defined by

$$D(T)x_k = t_{kk}x_k \ ; k = 1, \ldots, n,$$

where x_k is a coordinate of a vector x in the standard basis. Therefore

$$\|R_\lambda(D(T))\|_0 \le \frac{1}{\min_{k=1,\ldots,n} |t_{kk} - \lambda|} = \rho^{-1}(T,\lambda).$$

Corollary 1.6.2 gives

$$R_\lambda(T) = R_\lambda(D) \prod_{2 \le k \le n} (I - \frac{V(T)\Delta P_k}{\lambda - \lambda_k(T)}).$$

This finishes the proof; the detailed verification of the stated result is left to the reader. $\square$

Put

$$\delta(A,\lambda) \equiv \min_{k=1,\ldots,n} |\lambda - a_{kk}|.$$

Lemma 4.4.5 *Each eigenvalue $\lambda_m(A)$ of the matrix $A = (a_{jk})$ satisfies the inequality*

$$1 \le q_T \delta^{-1}(A, \lambda_m(A)) \prod_{k=2}^{n} (1 + \frac{m(k,A)}{|a_{kk} - \lambda_m(A)|}).$$

Proof: We use Lemma 4.1.3 taking $C = I, B = A$ and $L = T$. Here T is the upper-triangular matrix with the elements of the matrix A:

$$T = (t_{jk}) \text{ with } t_{jk} = a_{jk} \text{ if } j \le k, \text{ and } t_{jk} = 0 \text{ if } j > k.$$

We have

$$\|A - T\|_0 = \max_{j=2,\ldots,n} \sum_{k=1}^{j-1} |a_{jk}| = q_T.$$

Lemma 4.4.3 and Lemma 4.1.3 imply for $\lambda \in \sigma(A)$

$$1 \le q_T \|R_\lambda(T)\| \le q_T \rho^{-1}(T,\lambda) \prod_{k=2}^{n} (1 + \frac{m(k,T)}{|a_{kk} - \lambda|}).$$

This inequality coincides with the assertion of the lemma because $m(k,T) = m(k,A)$ and

$$\delta(A, \lambda_m(A)) = \rho(T, \lambda_m(A))$$

in this case. $\square$

Proof of Theorem 4.4.1: Since

$$|a_{kk} - \lambda| \geq \delta(A, \lambda), \text{ for all } k = 1, ..., n,$$

Lemma 4.4.5 implies the inequality

$$1 \leq q_T \delta^{-1}(A, \lambda) \prod_{k=2}^{n} (1 + m(k, A)\delta^{-1}(A, \lambda)) \tag{4.3}$$

for any eigenvalues λ of A. Dividing the algebraic equation (4.1) by z^n we can write down

$$1 = q_T z^{-1} \prod_{k=2}^{n} (1 + m(k, A)z^{-1}).$$

Comparing this with (4.3) and taking into account that $z(A)$ is the extreme right root of (4.1) we obtain:

$$\delta(A, \lambda) \leq z(A).$$

As claimed. □

4.4.3 Remarks

1. Consider a matrix of the second order: $A = (a_{jk}), j, k = 1, 2$. In this case $q_T = |a_{21}|, m(2, A) = |a_{12}|$, and $z(A)$ is the right-hand root of the equation

$$z^2 = |a_{21}|(z + |a_{12}|).$$

I.e.

$$z(A) = \frac{|a_{21}|}{2} + \sqrt{\frac{|a_{21}|^2}{4} + |a_{21}||a_{12}|}.$$

Theorem 4.4.1 implies the estimates

$$|a_{jj} - \lambda_j(A)| \leq z(A) \text{ for } j = 1, 2. \tag{4.4}$$

Gerschgorin's theorem (see e.g. (Stewart and Sun, 1990)) gives

$$|a_{11} - \lambda_1(A)| \leq |a_{12}| \tag{4.5}$$

and

$$|a_{22} - \lambda_2(A)| \le |a_{21}|.$$

For instance, let the condition

$$|a_{21}| \le \frac{|a_{12}|}{2} \tag{4.6}$$

hold. Then

$$z(A) = \frac{|a_{21}|}{2} + \sqrt{\frac{|a_{21}|^2}{4} + |a_{21}||a_{12}|} \le b|a_{12}|,$$

where

$$b = 1/4 + \sqrt{1/16 + 1/2} < 1.$$

That is, under (4.6) the inequality (4.4) with $j = 1$ is better than (4.5).

Let now n be arbitrary. In the case of an upper-triangular matrix A we have $q = 0$. Therefore, Theorem 4.4.1 implies the precise result

$$|a_{jj} - \lambda_j(A)| = 0 \text{ for all } j.$$

At the same time, Gerschgorin's result gives

$$|a_{jj} - \lambda_j(A)| \le \sum_{k=j+1}^{n} |a_{jk}| \text{ for all } j.$$

Thus, Theorem 4.4.1 improves Gerschgorin's theory for matrices which are "near" to triangular ones.

2. Again consider a matrix of the second order. Corollary 4.4.2 asserts that the condition

$$|a_{jj}| > z(A) = \frac{|a_{21}|}{2} + \sqrt{\frac{|a_{21}|^2}{4} + |a_{21}||a_{12}|} \tag{4.7}$$

guarantees the invertibility of the matrix. At the same time by the Levy-Desplanques theorem (Markus and Minc,1964, p.146) the following conditions:

$$|a_{11}| > |a_{12}| \text{ and } |a_{22}| > |a_{21}|. \tag{4.8}$$

are sufficient for non-singularity of A.

Let the condition (4.6) hold. Then (4.7) gives the following invertibility condition

$$|a_{jj}| > b|a_{12}| \ (j = 1, 2),$$

where b is the same as above.

That is, under (4.6) the condition (4.7) is better than (4.8).

In the case of an upper-triangular matrix A we have $q = 0$. Therefore, Corollary 4.4.2 implies the precise invertibility condition

$$a_{jj} \neq 0 \text{ for all } j.$$

At the same time, the Levy-Desplanques theorem gives the following non-singularity condition

$$|a_{jj}| > \sum_{k=j+1}^{n} |a_{jk}| \text{ for all } j.$$

Thus, Corollary 4.4.2 improves the Levy-Desplanques theorem for matrices which are "near" to triangular ones.

4.5 Bounds for the determinant of a matrix

Again, let A and B be $n \times n$-matrices,

$$q = \|A - B\|,$$

and

$$\chi(z) = max\{|z|, 1\}$$

for a complex number z, and

$$G_n(A) = \sum_{k=0}^{n-1} g^k(A)\gamma_{n,k}.$$

Recall that $g(A)$ and $\gamma_{n,k}$ are defined in Section 1.1.

Theorem 4.5.1 *Let an eigenvalue $\lambda_j(B)$ of B satisfy the inequality*

$$|\lambda_k(A) - \lambda_j(B)| \leq 1, \text{ for all } k = 1, ..., n. \tag{5.1}$$

Then the relation

$$\prod_{k=1}^{n} |\lambda_k(A) - \lambda_j(B)| \leq qG_n(A) \tag{5.2}$$

is true for all $j = 1, \ldots, n$.

Proof: Due to Lemma 4.1.3 we have

$$1 \leq q\|R_{\lambda_j(B)}(A)\|. \tag{5.3}$$

It is clear that under (5.1)

$$\chi(\lambda_k(A) - \lambda_j(B)) = 1.$$

Thus

$$\omega(A - I\lambda_j(B)) = 1.$$

Recall that

$$\omega(A - I\lambda_j(B)) = \prod_{j=1}^{n} \chi(A - I\lambda_j(B)).$$

Theorem 1.7.1 and (5.3) imply

$$1 \leq q|det(A - I\lambda_j(B))|^{-1}G_n(A).$$

As claimed.□

Corollary 4.5.2 *Let A and B be arbitrary $n \times n$ matrices. Then all eigenvalues of B lie in the set*

$$\{z \in \mathbf{C} : \prod_{k=1}^{n} |\lambda_k(A) - z| \leq qG_n(\frac{1}{M}A)M^{n-1}\},$$

where

$$M = \|A\| + \|B\|.$$

Indeed, put

$$A_1 = \frac{1}{M}A \text{ and } B_1 = \frac{1}{M}B.$$

Then

$$\|A_1 - B_1\| = \frac{q}{M},$$

and

$$|\lambda_k(A_1) - \lambda_j(B_1)| \le 1, \; j, k = 1, ..., n,$$

because

$$|\lambda_k(B_1)| = |\lambda_k(B)|/M \le \|B\|/M$$

and

$$|\lambda_k(A_1)| = |\lambda_k(A)|/M \le \|A\|/M.$$

Now Theorem 4.5.1 implies

$$\prod_{k=1}^{n}(|\lambda_k(A) - \lambda_j(B)|M^{-1}) \le qG_n(\frac{1}{M}A)M^{-1}.$$

or

$$\prod_{k=1}^{n}|\lambda_k(A) - \lambda_j(B)| \le qG_n(\frac{1}{M}A)M^{n-1}.$$

This inequality is equivalent to the stated result.

4.6 Approximations of a compact operator by matrices

4.6.1 Approximation of eigenvalues

Let H be a separable Hilbert space with the norm $\| \cdot \|$, K a linear compact operator acting in H, and A an operator generated by an $n \times n$-matrix

$$B = (b_{jk})_{j,k=1}^{n} \; ; n < \infty.$$

Namely, for some orthonormed basis $\{d_k\}$ we have

$$(Ad_k, d_j) = b_{jk} \text{ for } j, k \le n$$

and

$$(Ad_k, d_j) = 0 \text{ for } j, k \ge n + 1.$$

Denote

$$q = \| A - K \|.$$

By n_1 the number of different eigenvalues of A is denoted.

Theorem 4.6.1 *The spectrum of the compact operator K lies in the set*

$$\cup_{j=0}^{n_1}\Omega_j, \tag{6.1}$$

where

$$\Omega_j = \{\lambda \in \mathbf{C} : |\lambda - \lambda_j(B)| \leq z(q, B)\}, j = 1, ..., n_1,$$

and

$$\Omega_0 = \{\lambda \in \mathbf{C} : |\lambda| \leq z(q, B)\},$$

where $z(q, B)$ is the extreme right-hand (unique positive and simple) root of the algebraic equation

$$z^n = q \sum_{k=0}^{n-1} g^k(B)\gamma_{n,k} z^{n-k-1}. \tag{6.2}$$

Recall that $g(A)$ and $\gamma_{n,k}$ are defined in Section 1.1.

Proof: Theorem 1.2.2 gives

$$\| R_\lambda(B) \| \leq \sum_{k=0}^{n-1} \frac{g^k(B)\gamma_{n,k}}{\rho^{k+1}(B, \lambda)}. \tag{6.3}$$

Since A is the operator in H with the finite dimensional range generated by the matrix B, there is an n-dimensional orthogonal projector Q such that $AQ = BQ$ and $A(I - Q) = 0$.

It is simple to see that

$$R_\lambda(A) = R_\lambda(A)Q + R_\lambda(A)(I - Q).$$

and

$$R_\lambda(A)(1 - Q) = -\frac{1}{\lambda}(I - Q).$$

Therefore,

$$\| R_\lambda(A)(I - Q) \| \leq \frac{1}{|\lambda|}.$$

This inequality and (6.3) imply

$$\| R_\lambda(A) \| \leq \max\{\sum_{k=0}^{n-1} \frac{g^k(B)\gamma_{n,k}}{\rho^{k+1}(B, \lambda)}, \frac{1}{|\lambda|}\}. \tag{6.4}$$

The spectrum of A is

$$\sigma(A) = \sigma(B) \cup 0, \tag{6.5}$$

and $(I - Q)H$ is the eigenspace of A corresponding to 0. This means that

$$\rho(B, \lambda) \geq \rho(A, \lambda) \ and \ |\lambda| \geq \rho(A, \lambda)$$

for any complex λ. These inequalities and (6.4) imply

$$\| R_\lambda(A) \| \leq \sum_{k=0}^{n-1} \frac{g^k(B)\gamma_{n,k}}{\rho^{k+1}(A, \lambda)} \tag{6.6}$$

for all regular λ.

This inequality shows that in the considered case we can reduce the equation (1.7) (by multiplying by z^n) to the equation (6.2). Now Lemma 4.1.4 entails the inequality $sv_A(K) \leq z(B, q)$. According to (6.5) this inequality is equivalent to the assertion. $\square$

Naturally, if $0 \in \sigma(B)$, then the set defined by (6.1) can be rewritten in the form

$$\{\lambda \in C : |\lambda - \lambda_j(B)| \leq z(q, B), j = 1, ..., n_1\}.$$

Further, using (3.4) and (3.5) we obtain

Corollary 4.6.2 *Let under the assumptions of Theorem 4.6.1, the inequality $qw(n) \leq g(B)$ hold. Then the spectrum of the compact operator K lies in the set*

$$\cup_{j=0}^n \Omega_j^-,$$

where

$$\Omega_j^- = \{\lambda \in C : |\lambda - \lambda_j(B)| \leq g^{1-1/n}(B) \, [qw(n)]^{1/n}\}, \ j = 1, ..., n_1$$

and

$$\Omega_0^- = \{\lambda \in C : |\lambda| \leq g^{1-1/n}(B)[qw(n)]^{1/n}\}.$$

If under the assumptions of Theorem 4.6.1 $qw(n) > g(A)$, then the spectrum of K lies in the set

$$\cup_{j=0}^n \Omega_j^+$$

where

$$\Omega_j^+ = \{\lambda \in \mathbf{C} : |\lambda - \lambda_j(B)| \le qw(n)\}, j = 1, ..., n_1$$

and

$$\Omega_0^+ = \{\lambda \in C : |\lambda| \le qw(n)\}.$$

4.6.2   Conservation of multiplicities

Let $\Omega(c, a)$ be the disc on the complex plane with the center $c \in \mathbf{C}$ and the radius a, K be a compact operator, B be an $n \times n$-matrix, and $q = \|K - B\|$, again. Set

$$G_n(x, B) \equiv \sum_{j=0}^{n-1} \gamma_{k,n} x^j g^j(B)$$

for a positive scalar variable x .

Theorem 4.6.3 *Let B have an eigenvalue $\lambda(B) \ne 0$ of the algebraic multiplicity ν such that the distance from $\lambda(B)$ to the set $(\sigma(B) \cup 0)/\lambda(B)$ is equal to d. Let, in addition, there be a positive number $a < d$ such that*

$$0 \le \frac{qaG_n^2(1/a, B)}{1 - qG_n(1/a, B)} < 1.$$

Then the compact operator K has in $\Omega(\lambda(A), a)$ eigenvalues whose total algebraic multiplicity is equal to ν.

Proof: As mentioned above, B generates in H the operator A whose spectrum is given by (6.5). Now the result immediately follows from Lemma 4.2.1 and the estimate (6.6). $\square$

4.7   Spectrum perturbations of Hilbert-Schmidt operators

4.7.1   Bounds for the spectrum of a perturbed operator

Theorem 4.7.1 *Let A be a Hilbert-Schmidt operator and B be a linear bounded operator in H. Then the following bound is true:*

$$sv_A(B) \equiv \sup_{\mu(B) \in \sigma(B)} \inf_{\lambda(A) \in \sigma(A)} |\mu(B) - \lambda(A)| \le z(q, A), \qquad (7.1)$$

where $z(q, A)$ is the extreme right-hand (unique positive and simple) root of
the equation

$$q \sum_{k=0}^{\infty} \frac{g^k(A)}{\sqrt{k!} z^{k+1}} = 1. \tag{7.2}$$

Here $q = \|A - B\|$, again.

Proof: Theorem 2.4.1 gives the inequality

$$\|R_\lambda(A)\| \leq \sum_{k=0}^{\infty} \frac{g^k(A)}{\sqrt{k!} \rho^{k+1}(A, \lambda)} \text{ for all regular } \lambda. \tag{7.3}$$

Now the result is due to Lemma 4.1.4. $\square$

From Theorem 4.3.1 and Definition 4.1.2 follows

Theorem 4.7.2 *Let both A and B be Hilbert-Schmidt operators. Then*

$$hd(A, B) \equiv \max\{sv_A(B), sv_B(A)\} \leq \max\{z(q, A), z(q, B)\},$$

where $z(q, A)$ and $z(q, B)$ are the extreme right-hand (unique positive and
simple) roots of (7.2) and the following equation

$$q \sum_{k=0}^{\infty} \frac{g^k(B)}{\sqrt{k!} z^{k+1}} = 1$$

respectively.

In order to obtain a bound for $z(q, A)$ we introduce for an integer $r \geq 1$
and a fixed $a \geq 0$ the following scalar-valued function of a non-negative
variable z

$$\theta_r(a, z) \equiv q \sum_{m=0}^{r-1} \sum_{k=0}^{\infty} \frac{a^{kr+m}}{z^{kr+m+1}\sqrt{k!}}, \quad r > 1,$$

$$\theta_1(a, z) \equiv q \sum_{k=0}^{\infty} \frac{a^k}{z^{k+1}\sqrt{k!}}. \tag{7.4}$$

Let us consider the equation

$$q\theta_r(a, z) = 1 \tag{7.5}$$

Set

$$\zeta_r = \sqrt{2} \sum_{m=1}^{r} (m/2re)^{m/2r}, \quad \text{for } r > 1, \text{ and } \zeta_1 = \sqrt{2}.$$

Lemma 4.7.3 *Under the condition*

$$q\zeta_r \leq a, \tag{7.6}$$

the extreme right-hand (unique positive and simple) root $z(a)$ of the equation (7.5) satisfies the inequality

$$z^{2r}(a) \leq \frac{2a^{2r}}{\ln a - \ln(q\zeta)}$$

Proof: First, let $r > 1$. Schwarz's inequality gives

$$\theta_r(a, z) \leq \sum_{m=0}^{r-1} \frac{a^m}{z^{m+1}} [\sum_{k=0}^{\infty} \frac{2^k a^{2kr}}{z^{2kr} k!} \sum_{j=0}^{\infty} 2^{-j}]^{1/2}.$$

That is,

$$\theta_r(a, z) \leq \sqrt{2} \sum_{m=0}^{r-1} \frac{a^m}{z^{m+1}} exp[a^{2r}/z^{2r}].$$

This inequality and (7.5) imply the relation

$$q\sqrt{2} \sum_{m=0}^{r-1} \frac{a^m}{z^{m+1}(a)} exp[a^{2r}/z^{2r}(a)] \geq 1. \tag{7.7}$$

Differentiating, we can show that

$$\max_{y \geq 0} y^k exp[-y^{2r}] = (k/2re)^{k/2r} \text{ for } 1 \leq k \leq 2r.$$

Thus,

$$(a/z)^k \leq exp[(a/z)^{2r}](k/2re)^{k/2r}.$$

Now (7.7) gives

$$1 \leq \frac{\sqrt{2}q}{a} \sum_{m=0}^{r-1} \frac{a^{m+1}}{z^{m+1}(a)} e^{[a/z(a)]^{2r}} \leq$$

$$\frac{\sqrt{2}q}{a} \sum_{m=1}^{r} (m/2re)^{m/2r} e^{2[a/z(a)]^{2r}}.$$

Or

$$1 \leq qa^{-1} \zeta_r e^{2[a/z(a)]^{2r}}.$$

This implies

$$ln(a/\zeta_r q) \leq 2(a/z(a))^{2r}.$$

As claimed. Similarly, by Schwarz's inequality the stated result is proved in the case $r = 1$. $\square$

For $a = g(A), r = 1$ the equations (7.2) and (7.5) coincide. Thus, under the condition

$$\sqrt{2}q \leq g(A) \tag{7.8}$$

by Lemma 4.7.3 we have the estimation

$$z(q, A) \leq \frac{\sqrt{2}g(A)}{\sqrt{\ln(g(A)/q\sqrt{2})}}.$$

This bound and Theorem 4.7.1 imply

Corollary 4.7.4 *Let A be a Hilbert-Schmidt operator and B be a linear bounded operator in H, and let the condition (7.8) hold. Then the inequality*

$$sv_A(B) \leq \frac{\sqrt{2}g(A)}{\sqrt{\ln(g(A)/q\sqrt{2})}}$$

is true.

Setting

$$m(A, B) = max\{g(A), g(B)\}$$

we derive by Theorem 4.7.2 the following

Corollary 4.7.5 *Let A and B be Hilbert-Schmidt operators, and let the conditions (7.8), and*

$$\sqrt{2}q \leq g(B)$$

hold. Then

$$h(A, B) \leq \frac{\sqrt{2}m(A, B)}{\sqrt{\ln[m(A, B)/\sqrt{2}q]}}.$$

4.7.2   Perturbation of multiplicities

Set

$$G(x, A) \equiv \sum_{j=0}^{\infty} \frac{x^j \, g^j(A)}{\sqrt{j!}} \tag{7.9}$$

for a positive scalar variable x .

Theorem 4.7.6 *Let A be a Hilbert-Schmidt operator and let B be a bounded operator in H. Suppose A has an eigenvalue $\lambda(A)$ of the algebraic multiplicity ν such that the distance from $\lambda(A)$ to the rest of $\sigma(A)$ is equal to d, i.e. (2.3) holds. If, in addition, there is a positive number $a \leq d$ such that the condition*

$$0 \leq \frac{qaG^2(1/a, A)}{1 - qG(1/a, A)} < 1$$

holds, then operator B has in $\Omega(\lambda(A), a)$ eigenvalues whose total algebraic multiplicity is equal to ν.

Proof: In fact, this result immediately follows from Lemma 4.2.1 and the estimate (7.3).□

4.8   A bound for the determinant of a nuclear operator

Recall that the function $G(x, A)$ is defined by (7.9).

Theorem 4.8.1 *Let A be a Hilbert-Schmidt operator with a finite trace, and let a point $\lambda(B)$ of the spectrum of a bounded operator B satisfy the inequality*

$$|1 - \frac{\lambda_k(A)}{\lambda(B)}| \leq 1, \text{ for all } k = 1, 2, \ldots. \tag{8.1}$$

Then the inequality

$$\prod_{k=1}^{\infty} |1 - \frac{\lambda_k(A)}{\lambda(B)}| \leq \frac{q}{|\lambda(B)|} G(\frac{1}{\lambda(B)}, A) \tag{8.2}$$

is valid.

Proof: According to Lemma 4.1.3

$$q\|(A - I\lambda(B))^{-1}\| \geq 1.$$

Since

$$1 \leq q\|(A - \lambda(B)I)^{-1}\| = \frac{1}{|\lambda(B)|}\|(A\lambda_j^{-1}(B) - I)^{-1}\|,$$

we obtain due to Theorem 2.9.1

$$1 \leq \frac{qG(\lambda^{-1}(B)A)}{|det(I - A\lambda^{-1}(B))\,\lambda(B)|},$$

because according to (8.1)

$$\chi(1 - \lambda_k(A)\lambda^{-1}(B)) = 1 \ \ (k = 1, ..., n),$$

and, therefore,

$$\omega(I \ \ A\lambda^{-1}(B)) = 1.$$

This proves the result. □

Corollary 4.8.2 *Let A be a normal nuclear operator, and* $\lambda(B)$ *be a point of the spectrum of a bounded operator B. Then under condition (8.1) eigenvalues of A the inequality*

$$\prod_{k=1}^{\infty} |1 - \lambda_k(A)\lambda^{-1}(B)| \leq \frac{q}{|\lambda(B)|}.$$

is valid.

Indeed, in this case

$$g(A) = 0, \ \ \text{and} \ \ G(A\lambda) = 1.$$

Now Theorem 4.8.1 ensures the result.

4.9   Spectrum perturbations of operators with Hilbert-Schmidt powers

4.9.1   Perturbations of eigenvalues

Let A and B be bounded linear operators acting in a separable Hilbert space H. Assume that the condition

$$A^p \text{ is a Hilbert-Schmidt operator} \qquad (9.1)$$

holds with some integer $p \geq 1$. By $n_1 \leq \infty$ the number of different eigenvalues of A is denoted, $q = \|B - A\|$, and

$$\mu_p(A, B) \equiv \sum_{k=0}^{p-1} \|A^k\| \, \|B^{p-k-1}\|.$$

To formulate the result we denote by $z_p(A, q)$ the extreme right-hand (unique positive and simple) root of the equation

$$1 = q\mu_p(A, B) \sum_{k=0}^{\infty} \frac{g^k(A^p)}{z^{k+1}\sqrt{k!}}. \qquad (9.2)$$

Theorem 4.9.1 *Let a linear operator A acting in H satisfy (9.1) for some positive integer p and let B be a linear bounded operator in H. Then the spectrum of B lies in the set*

$$\cup_{j=1}^{n_1} \Omega_j,$$

where

$$\Omega_k = \{z \in \mathbf{C} : |z^p - \lambda_k^p(A)| \leq z_p(A, q)\}; \; k = 1, 2, ..., n_1. \qquad (9.3)$$

Proof: Due to Theorem 2.5.1

$$\|R_\lambda(A)\| \leq \|T_{\lambda,p}\| \sum_{k=0}^{\infty} \frac{g^k(A^p)}{\rho^{k+1}(A^p, \lambda^p)(k!)^{1/2}}, \qquad (9.4)$$

where

$$T_{\lambda,p} = \sum_{k=0}^{p-1} A^k \lambda^{p-k-1},$$

and

$$\rho(A^p, \lambda^p) = \inf_{t \in \sigma(A)} |t^p - \lambda^p|$$

is the distance between $\sigma(A^p)$ and the point λ^p. But

$$|\lambda^j| \le \|B^j\| \text{ for } \lambda \in \sigma(B).$$

Thus,

$$\|T_{\lambda,p}\| \le \sum_{k=0}^{p-1} \|A^k\| \|B^{p-k-1}\| = \mu_p(A, B).$$

Hence,

$$\|R_\lambda(A)\| \le \mu_p(A, B) \sum_{k=0}^{\infty} \frac{g^k(A^p)}{\rho^{k+1}(A^p, \lambda^p)(k!)^{1/2}}. \tag{9.5}$$

for $\lambda \in \sigma(B)$. Now Lemma 4.1.3 implies

$$1 \le q \|R_\lambda(A)\| \text{ for any } \lambda \in \sigma(B).$$

Combining this inequality with (9.5) we deduce the inequality

$$1 \le q\mu_p(A, B) \sum_{k=0}^{\infty} \frac{g^k(A^p)}{\rho^{k+1}(A^p, \lambda^p)\sqrt{k!}}$$

for all $\lambda \in \sigma(B)$. Comparing this inequality with (9.2) and taking into account that $z_p(A, q)$ is the root of (9.2), we arrive at the stated result. $\square$

The following result is due to Lemma 4.7.3 and Theorem 4.9.1

Corollary 4.9.2 *Let a linear operator A acting in H satisfy (9.1) for some positive integer p and let B be a linear bounded operator in H. Let, in addition,*

$$\sqrt{2}q\mu_p \le g(A^p).$$

Then the spectrum of B lies in the set

$$\cup_{j=1}^{n_1} \Omega_j^+,$$

where

$$\Omega_k^+ = \{z \in \mathbf{C} : |z^p - \lambda_k^p(A)| \le z^+\}, \; k = 1, 2, ..., n_1,$$

with the notation

$$z^+ = \frac{\sqrt{2}g(A^p)}{\sqrt{\ln g(A^p) - \ln[\sqrt{2}q\mu_p(A,B)]}}.$$

4.9.2   Perturbations of multiplicities of operators with Hilbert-Schmidt powers

We need the following:

Lemma 4.9.3 *Let A and B be linear operators in H with the property $q = \|A - B\| < \infty$. Further, let A have an eigenvalue $\lambda(A)$ of the algebraic multiplicity ν and the distance from $\lambda(A)$ to the rest of $\sigma(A)$ is equal to d, i.e. (2.3) holds, and for a positive number $a < d$, the following relation holds:*

$$0 \le \frac{qa\psi^2(a)}{1 - q\psi(a)} < 1 \tag{9.6}$$

with the notation

$$\psi(a) = \sup_{|\lambda(A) - \lambda| = a} \| R_\lambda(A) \|.$$

Then the operator B has in $\Omega(\lambda(A), a)$ eigenvalues whose total algebraic multiplicity is equal to ν.

The proof of this lemma repeats almost word for word the proof of Lemma 4.2.1

For $a < d$ and $|\lambda(A)| > a$ under (9.1) by (9.4) we have

$$\sup_{|\lambda(A) - \lambda| = a} \|R_\lambda(A)\| \le G(a, A^p) \equiv$$

$$\sum_{k=0}^{p-1} \|A^k\|(|\lambda(A)| + a)^{p-k-1} \sum_{k=0}^{\infty} \frac{g^k(A^p)}{(|\lambda(A)|^p - a^p)^{k+1}\sqrt{k!}}. \tag{9.7}$$

This estimate and Lemma 4.9.3 yield

Theorem 4.9.4 *Let A and B be a bounded operator in H, and A satisfy (9.1). Further, suppose A has an eigenvalue $\lambda(A)$ of the algebraic multiplicity ν such that the distance from $\lambda(A)$ to the rest of $\sigma(A)$ is equal to d,* *i.e. (2.3) holds. If in addition, there is a positive number $a \leq d$ such that the condition*

$$0 \leq \frac{qaG^2(1/a, A^p)}{1 - qG(1/a, A^p)} < 1$$

holds, then operator B has in $\Omega(\lambda(A), a)$ eigenvalues whose total algebraic multiplicity is equal to ν.

Note that the inequality $|\lambda(A)| \geq d$ is satisfied, since obviously $0 \in \sigma(A)$. So the condition $|\lambda(A)| > a$ must be automatically fulfilled, because $a < d$.

4.10 Spectrum perturbations of Neumann-Schatten operators

4.10.1 Perturbations of eigenvalues

In this section we analyze spectrum perturbations of a linear operator A acting in a separable Hilbert space H and satisfying the condition

$$A \in C_{2r}, \text{ for some integer } r, \tag{10.1}$$

where C_{2r} is the Neumann-Schatten's ideal (see Section 2.1). Let for a fixed $b \geq 0$

$$\theta_r(b, z) = \sum_{m=0}^{r-1} \sum_{k=0}^{\infty} \frac{b^{kr+m}}{z^{rk+m+1}\sqrt{k!}} \text{ for } z > 0.$$

Theorem 4.10.1 *Let a linear operator A acting in H satisfy (10.1) and B be a bounded linear operator in H. Then the inequality*

$$sv_A(B) \equiv \sup_{\mu(B)\in\sigma(B)} \inf_{\lambda(A)\in\sigma(A)} |\mu(B) - \lambda(A)| \leq z(q, A) \tag{10.2}$$

is valid, where $z(q, A)$ is the extreme right-hand (unique positive and simple) root of the scalar equation

$$1 = q\theta_r(2N_{2r}(A), z).$$

Proof: Theorem 2.6.1 gives the inequality

$$\|R_\lambda(A)\| \le \theta_r(2N_{2r}(A), \rho(A, \lambda)) \text{ for all regular } \lambda \qquad (10.3).$$

Now the result is due to Lemma 4.1.4 □

   The following result is easily derived with the help Lemma 4.7.3 and Theorem 4.10.1:

Corollary 4.10.2 *Let (10.1) be satisfied and B be a linear bounded operator in H. Let the condition*

$$q\zeta_r \le 2N_{2r}(A) \qquad (10.4)$$

hold. Then the following inequality is true:

$$sv_A(B) \le \frac{2 \sqrt[2r]{2} N_{2r}(A)}{\sqrt[2r]{\ln(2N_{2r}(A)/q\zeta_r)}}.$$

   From Theorem 4.10.1 and Definition 4.1.2 it follows

Theorem 4.10.3 *Let A satisfy (10.1) and let a linear operator B satisfy the condition*

$$B \in C_{2l} \text{ for some integer positive } l.$$

Then

$$hd(A, B) \le \max\{z(q, A), z(q, B)\},$$

where $z(q, B)$ is the extreme right-hand (unique positive and simple) root of the equation $q\theta_l(2N_{2l}(B), z) = 1$.

4.10.2   Perturbations of multiplicities

Theorem 4.10.4 *Let A satisfy (10.1) and let B be a bounded operator in H. Suppose A has an eigenvalue $\lambda(A)$ of the algebraic multiplicity ν such that the distance from $\lambda(A)$ to the rest of $\sigma(A)$ is equal to d, i.e. (2.3) holds. If in addition, there is a positive number $a \le d$ such that the condition*

$$0 \le \frac{qa\theta_{2r}^2(2N_{2r}(A), a)}{1 - q\theta_{2r}(2N_{2r}(A), a)} < 1$$

holds, then operator B has in $\Omega(\lambda(A), a)$ eigenvalues whose total algebraic multiplicity is equal to ν.

Proof: This result follows from Lemma 4.2.1 and the estimate (10.3).$\square$

4.11 Bounds for eigenvalues of nuclear operators

4.11.1 Statement of the result

In this subsection we derive bounds for eigenvalues of Trace class operators under certain conditions below.

Let a completely continuous operator A in H be represented by an infinite matrix with the entries a_{jk}; $j, k = 1, 2, \ldots$. Put

$$m(k, A) = \max_{j=1,\ldots,k-1} |a_{jk}| \quad \text{for} \quad k = 2, 3, \ldots, .$$

Let us suppose that

$$q_T \equiv \max_{j=2,3,\ldots} \sum_{k=1}^{j-1} |a_{jk}| < \infty \tag{11.1}$$

and

$$\nu(A) \equiv \sum_{k=1}^{\infty} m(k, A) < \infty. \tag{11.2}$$

Theorem 4.11.1 *Let (11.1) and (11.2) hold. Then all the eigenvalues of the operator A lie in the set*

$$\cup_{k=2}^{\infty} \Omega(a_{kk}, z(A)),$$

where $\Omega(a_{kk}, z(A)) = \{\lambda \in \mathbf{C} : |\lambda - a_{kk}| \leq z(A)\}$; $k = 1, 2, \ldots$, and $z(A)$ is the extreme right-hand root of the equation

$$1 = \frac{q_T}{z} \prod_{k=2}^{\infty} (1 + \frac{m(k, A)}{z}). \tag{11.3}$$

Proof: Let $\{A_n\}$ be a sequence of linear n-dimensional operators represented by the $n \times n$-matrices (a_{jk}). It is simple to check that this sequence converges to A in the operator norm. Now one can easily obtain the stated result by applying Theorem 4.4.1 and Lemma 2.4.2. The details are left to the reader. □

Lemma 4.11.2 *Let the condition*

$$q_T \geq e\nu(A) \tag{11.4}$$

hold. Then the extreme right root $z(A)$ of (11.3) satisfies the estimate

$$z(A) \leq z_+ \equiv \frac{2\nu(A)}{\ln(e\nu(A)/q_T)}. \tag{11.5}$$

Proof: Since $1 + a \leq exp(a)$ for $a \geq -1$, we can write down by (11.3) $1 \leq q_T z^{-1} exp[\nu(A)z^{-1}]$, $z = z(A)$. Take into account that

$$\frac{q_T}{z} = \frac{q_T}{\nu(A)}(\frac{\nu(A)}{z} - 1 + 1) \leq \frac{q_T}{\nu(A)} exp[\frac{\nu(A)}{z} - 1].$$

Thus,

$$1 \leq \frac{q_T}{\nu(A)} exp[\frac{2\nu(A)}{z(A)} - 1].$$

Hence,

$$\frac{2\nu(A)}{z(A)} - 1 \geq \ln(\nu(A)/q_T),$$

Now the stated result is due to (11.4). □

This lemma and Theorem 4.11.1 entail the following

Corollary 4.11.3 . *Let (11.1), (11.2), and (11.4) hold. Then all the eigenvalues of A lie in the discs $\{\lambda \in \mathbf{C} : |\lambda - a_{kk}| \leq z_+\}$; $k = 1, 2, ...$.*

4.12 Notes

The spectrum perturbation theory has two branches. In this book we consider comparatively unstructured errors and attempt to bound the perturbations in terms of norms of the errors. Other approaches impose some structure on the errors: for example they may be analytic functions (see for instance Kato (1966), Baumgartel (1985)).

In the case of matrices we follow the book by Stewart and Sun (1990). Particular mention should be made of the very interesting book by Bhatia (1987).

The contents of **Section 4.1** are taken from Gil' (1993c) while the results of **Section 4.2** are adapted from Gil' (1991b).

The material of **Section 4.3** is taken from Gil' (1991a), Gil' (1993c).

Notice that very strong results concerning the perturbations of eigenvalues of matrices were obtained in Phillips (1990), Bhatia, Elsner, and Krause (1990).

The contents of **Sections 4.4, 4.5, and 4.6** are new. As mentioned above, Theorem 4.4.1 improves Gerschgorin's theorem for matrices which are near to triangular ones.

The material of **Sections 4.7 and 4.8** is based on Gil' (1991a), Gil' (1992), and Gil' (1993c) The contents of Sections **4.9, 4.10, and 4.11** are new.

Chapter 5

Perturbations of Noncompact Operators

This chapter is devoted to spectrum perturbations of noncompact operators.

In Section 5.1 spectrum perturbations of quasi-Hermitian generally unbounded operators are considered.

In Section 5.2 perturbations of quasiunitary operators are investigated.

Section 5.3 is devoted to spectrum perturbations of unbounded operators with compact resolvents.

5.1 Perturbations of quasi-Hermitian operators

Recall that

$$\beta_r = 2(1 + \frac{2r}{exp(2/3)ln2})$$

and

$$\theta_r(a, z) \equiv \sum_{m=0}^{r-1} \sum_{k=0}^{\infty} \frac{a^{kr+m}}{z^{kr+m+1}\sqrt{k!}}$$

163

for non-negative a, z, and an integer r.

In this section we analyze spectrum perturbations of a quasi-Hermitian operator A acting in a separable Hilbert space H and satisfying the condition

$$A_I \in C_{2r}, \text{ for some positive integer } r. \tag{1.1}$$

Theorem 5.1.1 *Let a generally unbounded quasi-Hermitian operator A acting in H satisfy the condition (1.1) and let B be a linear operator in H with the property*

$$q \equiv \|A - B\| < \infty. \tag{1.2}$$

Then

$$sv_A(B) \equiv \sup_{\mu(B) \in \sigma(B)} \inf_{\lambda(A) \in \sigma(A)} |\mu(B) - \lambda(A)| \le z_r(q, A),$$

where $z_r(q, A)$ is the extreme right-hand (unique positive and simple) root of the equation

$$q\theta_r(\beta_r N_{2r}(A), z) = 1.$$

Proof: Theorem 3.4.8 gives the inequality

$$\|R_\lambda(A)\| \le \theta_r(\beta_r N_{2r}(A), \rho(A, \lambda)) \text{ for all regular } \lambda. \tag{1.3}$$

Now the result is due to Lemma 4.1.4.□

From Theorem 5.1.1 and Definition 4.1.2 follows

Corollary 5.1.2 *Let linear operators A and B satisfy the conditions (1.1), (1.2), and*

$$B_I = (B - B^*)/2i \in C_{2l}.$$

with a positive integer l. Then

$$hd(A, B) \equiv \max\{sv_A(B), sv_B(A)\} \le \max\{z_r(q, A), z_l(q, B)\},$$

where $z_l(q, B)$ is the extreme right-hand (unique positive and simple) root of the equation

$$q\theta_l(\beta_l N_{2l}(B_I), z) = 1.$$

Note that Lemma 4.7.3 gives us the estimation for the root $z_r(q, A)$.

Now we make Theorem 5.1.1 more precise in the case of a (generally unbounded) quasi-Hermitian operator A acting in H and satisfying

$$A_I \in C_2. \tag{1.4}$$

Recall that $g_1(A)$ is defined in Section 3.4.

Theorem 5.1.3 *Let (1.2) and (1.4) hold. Then*

$$sv_A(B) \leq z_0(q, A),$$

where $z_0(q, A)$ is the extreme right-hand (unique positive and simple) root of the equation

$$q\theta_1(g_1(A), z) \equiv q \sum_{k=1}^{\infty} \frac{g_1^k(A)}{z^{k+1}\sqrt{k!}} = 1.$$

Proof: Theorem 3.4.2 gives the inequality

$$\|R_\lambda(A)\| \leq \theta_1(g_1(A), \rho(A, \lambda)) \text{ for all regular } \lambda.$$

Now, the result is due to Lemma 4.1.4. □

Put

$$m_0(A, B) = max\{g_1(A), g_1(B)\}.$$

Theorem 5.1.3 and Definition 4.1.2 imply:

Corollary 5.1.4 *Let the conditions (1.2) and (1.4) be fulfilled and and let B be quasi-Hermitian with the property*

$$B_I = (B - B^*)/2i \in C_2.$$

Then

$$hd(A, B) \leq z_0(q, A, B)),$$

where $z_0(q, A, B)$ is the extreme right-hand (unique positive and simple) root of the equation

$$q\theta_1(m_0(A, B), z) = 1.$$

In particular, let A be a selfadjoint operator. Then $g_1(A) = 0$, $z_0(q, A) = q$ and Theorem 5.1.3 gives the inequality $sv_A(B) \leq q$. Similarly, if both operators A and B are selfadjoint, then Theorem 5.1.3 implies the inequality $hd(A, B) \leq q$. Therefore, Theorem 5.1.3 generalizes the well-known result for spectrum perturbations of selfadjoint operators (Kato, 1966, p.291).

Note that Lemma 4.7.3 gves us an estimate for the roots $z_0(q, A)$ and $z_0(q, A, B)$

5.2 Perturbations of quasiunitary operators

Let a linear operator A in H have the property

$$A^*A - I \in C_1. \tag{2.1}$$

Recall that $\vartheta(A)$ is defined in Section 3.5.

Theorem 5.2.1 *Let (2.1) hold and let B be a bounded linear operator. Then*

$$sv_A(B) \leq x_0(q, A),$$

where $x_0(q, A)$ is the extreme right-hand (unique positive and simple) root of the equation

$$q\theta_1(\vartheta(A), z) = 1. \tag{2.2}$$

Here $q = \|A - B\|$.

Proof: Theorem 3.5.1 gives the inequality

$$\|R_\lambda(A)\| \leq \theta_1(\vartheta(A), \rho(A, \lambda)) \text{ for all regular } \lambda$$

Now, the result is due to Lemma 4.1.4. $\square$

Put

$$m_2(A, B) = max\{\vartheta(A), \vartheta(B)\}.$$

Theorem 5.2.1 and Definition 4.1.2 imply:

Corollary 5.2.2 *Let the conditions (2.1) and*

$$BB^* - I \in C_1$$

be fulfilled. Then

$$hd(A, B) \leq x_0(q, A, B),$$

where $x_0(q, A, B)$ is the extreme right-hand (unique positive and simple) root of the equation

$$q\theta_1(m_2(A, B), z) = 1.$$

5.3 Operators with compact resolvents

5.3.1 Statement of the result

Let H be a separable Hilbert space with the norm $\|.\|$, again. Throughout this section A is a linear generally unbounded invertible operator in H, and B is a linear operator with the property

$$q_0 \equiv \|(A - B)A^{-1}\| < 1. \tag{3.1}$$

Besides, we assume that for some natural number r

$$A^{-1} \in C_{2r}. \tag{3.2}$$

The aim of this section is to prove the following

Theorem 5.3.1 *Let the conditions (3.1) and (3.2) be satisfied. Then for any point $\lambda(B) \in \sigma(B)$, there is a point $\lambda(A) \in \sigma(A)$ such that the following inequality holds:*

$$\left| \frac{1}{\lambda(B)} - \frac{1}{\lambda(A)} \right| \leq y_r(q, A), \tag{3.3}$$

where $y_r(q, A)$ is the extreme right (unique positive and simple) root of the scalar equation

$$q\theta_r(2N_{2r}(A^{-1}), z) = 1 \text{ with } q = \frac{\|A^{-1}\|q_0}{1 - q_0}. \tag{3.4}$$

The proof is given in the next subsection.

Recall that

$$\zeta_r = \sqrt{2} \sum_{k=1}^{r} (k/2re)^{k/2r}.$$

Thanks to Lemma 4.7.3 we can assert that under the condition

$$q\zeta_r \leq 2N_{2r}(A^{-1}), \tag{3.5}$$

the inequality

$$y_r^{2r}(q, A) \leq M_r(q, A) \tag{3.6}$$

is true. Here

$$M_r(q, A) \equiv \frac{2^{2r+1} N_{2r}^{2r}(A^{-1})}{ln(2N_r(A^{-1})) - ln(q\zeta_r)}.$$

Corollary 5.3.2 *Let the conditions (3.1) and (3.2) be satisfied and let (3.5) hold. Then for any point $\lambda(B) \in \sigma(B)$ there is a point $\lambda(A) \in \sigma(A)$ such that the inequality*

$$|\frac{1}{\lambda(B)} - \frac{1}{\lambda(A)}| \leq \sqrt[2r]{M_r(q, A)}$$

is valid.

5.3.2   Proofs

We recall that due to Theorem 2.6.1 for any $K \in C_{2r}$

$$\|R_\lambda(K)\| \leq \theta_r(2N_{2r}(K), \rho(K, \lambda)) =$$

$$\sum_{m=0}^{r-1} \sum_{k=0}^{\infty} \frac{(2N_{2r}(K))^{kr+m}}{\rho^{kr+m+1}(K, \lambda)\sqrt{k!}}. \tag{3.7}$$

Lemma 5.3.3 *Let A and B be linear operators in a generally nonseparable Hilbert space and let the condition (3.1) be satisfied. Let in addition the inequality*

$$\|(\lambda - A^{-1})^{-1}\| \leq \psi(\rho(A^{-1}, \lambda))$$

hold, where $\psi(x)$ is a monotonically decreasing non-negative function of a positive variable x, and

$$\rho(A^{-1}, \lambda) = \inf_{\tau \in \sigma(A^{-1})} |\lambda - \tau|.$$

Then for any point $\lambda(B) \in \sigma(B)$ there is a point $\lambda(A) \in \sigma(A)$ such that

$$\left| \frac{1}{\lambda(B)} - \frac{1}{\lambda(A)} \right| \leq z(\psi).$$

Here $z(\psi)$ is the extreme right (unique positive and simple) root of the scalar equation

$$q\psi(z) = 1. \tag{3.8}$$

Proof: Since

$$B^{-1} - A^{-1} = B^{-1}(A - B)A^{-1},$$

we obtain

$$\|B^{-1} - A^{-1}\| \leq \|B^{-1}\| \, q_0.$$

This implies according to (3.1) the inequalities

$$\|B^{-1}\| \leq \|A^{-1}\| (1 - q_0)^{-1} \text{ and } \|B^{-1} - A^{-1}\| \leq q = \|A^{-1}\| q_0 (1 - q_0)^{-1}.$$

We invoke Lemma 4.1.4. It implies $sv_{A^{-1}}(B^{-1}) \leq z(\psi)$. Thanks to the spectral mapping theorem (see e.g.(Dunford and Schwartz, 1958)), this result is equal to the assertion. $\square$

Proof of Theorem 5.3.1: Due to (3.7) we have

$$\|(A^{-1} - \lambda)^{-1}\| \leq \theta_r(2N_{2r}(A^{-1}), \rho(A^{-1}, \lambda)).$$

Now Lemma 5.3.3 ensures the result. $\square$

5.3.3 Normal operators

In this subsection we consider the case of normal operators. Condition (3.2) is not assumed. Let A be a normal operator. Then it can be represented in the form

$$A = \int_{\sigma(A)} t \, dE_t,$$

where E_t is the spectral function of A (see e.g. Dunford and Schwartz (1963)). By this representation we easily obtain the estimation

$$\|(A^{-1} - I\lambda)^{-1}\| \leq [\inf_{t \in \sigma(A)} |\lambda - \frac{1}{t}|]^{-1} = \rho^{-1}(A^{-1}, \lambda)$$

(see also Kato (1966, p. 277)). Now Lemma 5.3.3 implies the following result:

Corollary 5.3.4 *Let A be a normal operator in a generally nonseparable Hilbert space and let the condition (3.1) be satisfied. Then for any point $\lambda(B) \in \sigma(B)$ there is a point $\lambda(A) \in \sigma(A)$ such that the inequality*

$$|\frac{1}{\lambda(B)} - \frac{1}{\lambda(A)}| \leq q$$

is valid.

5.3.4 Operators with Hilbert-Schmidt's resolvents

Let

$$A^{-1} \in C_2 \tag{3.9}$$

i.e. A^{-1} is a Hilbert-Schmidt operator. In order to obtain the result we employ Theorem 2.4.1. It asserts that for any operator $K \in C_2$, the relation

$$\|R_\lambda(K)\| \leq \theta_1(g(K), \rho(K, \lambda)) =$$

$$\sum_{k=0}^{\infty} \frac{g^k(K))}{\rho^{k+1}(K, \lambda)\sqrt{k!}} \tag{3.10}$$

is valid, where

$$g(K) = [N_2^2(K) - \sum_{k=0}^{\infty} |\lambda_k(K)|^2]^{1/2}.$$

Using (3.10) and omitting the simple calculations, we obtain

$$\|(A^{-1} - \lambda)^{-1}\| \leq \theta_1(g(A^{-1}), \rho(A^{-1}, \lambda)).$$

Now Lemma 5.3.3 gives the following

Corollary 5.3.5 *Let (3.1) and (3.9) hold. Then for any point* $\lambda(B) \in \sigma(B)$ *there is* $\lambda(A) \in \sigma(A)$ *such that*

$$\left| \frac{1}{\lambda(B)} - \frac{1}{\lambda(A)} \right| \le z_2,$$

where z_2 *is the extreme right (unique positive and simple) root of the equation*

$$q\theta_1(g(A^{-1}), z) = 1.$$

This result enables us to make Theorem 5.3.1 more precise under (3.9), since we replace $2N_2(A^{-1})$ by $g(A^{-1})$. Similarly, we can do it in the assertion of Corollary 5.3.2.

5.4 Notes

The contents of Sections 5.1 and 5.2 are based on Gil' (1991b) and Gil' (1992).

As mentioned above, Theorem 5.1.3 generalizes the well-known result for perturbations of selfadjoint operators (Kato, 1966, p.291). The results of Section 5.3 are new.

Chapter 6

Perturbations of Operators on a Tensor Product of Hilbert Spaces

This chapter is devoted to relatively bounded perturbations of unbounded operators defined on a tensor product of Hilbert spaces.

In Sections 6.1, 6.2, and 6.3 a class of operator-valued functions on a tensor product of Hilbert spaces is introduced. For the operator-valued functions, norm estimations are obtained.

In Sections 6.4 and 6.5 spectrum perturbations of unbounded operators on the tensor product are investigated by the norm estimations for the operator-valued functions.

In Section 6.6 some well-known results about differential operators are collected.

In Section 6.7 a bound is obtained for eigenvalues of a second order nonselfadjoint matrix differential operator on a segment.

The spectrum of a periodic matrix differential operator is investigated in Section 6.8.

In Section 6.9 a bound is established for the spectrum of a high order matrix differential operator on the real axis.

6.1   Tensor-valued functions

6.1.1   Notation

From the theory of tensor products we only need the basic definition and elementary facts which are to be found in Dunford and Schwartz (1958).

Let H be a tensor product of a Euclidean space $\mathbf{C}^n$ and a separable Hilbert space L, i.e. $H = \mathbf{C}^n \otimes L$. This means that H is the collection of all formal sums of the form

$$u = \sum_j y_j \otimes h_j \; ; y_j \in L, h_j \in \mathbf{C}^n \tag{1.1}$$

with the understanding that

$$\lambda(y \otimes h) = (\lambda y) \otimes h = y \otimes (\lambda h),$$
$$(y + y_1) \otimes h = y \otimes h_j + y_1 \otimes h,$$
$$y \otimes (h + h_1) = y \otimes h + y \otimes h_1.$$

Here $y, y_1 \in L \; ; h, h_1 \in \mathbf{C}^n$, and λ is a number.

The scalar product in H is defined as

$$(y \otimes h, y_1 \otimes h_1)_H = (y, y_1)_L \, (h, h_1)_{\mathbf{C}^n} \; ;$$

$$\text{with } y, y_1 \in L, \text{ and } h, h_1 \in \mathbf{C}^n$$

where $(.,.)_L$ and $(.,.)_{\mathbf{C}^n}$ are the scalar products in L and $\mathbf{C}^n$, respectively.

The norms in L, $\mathbf{C}^n$ and H are denoted by $\|.\|_L$, $\|.\|_{\mathbf{C}^n}$ and $\|.\|_H$, respectively.

If in (1.1) y_j are mutually orthogonal, then.

$$\|u\|_H = \sqrt{\sum_j \|y_j\|_L^2 \cdot \|h_j\|_{\mathbf{C}^n}^2} \tag{1.2}$$

Recall that by $D(A)$ and $\sigma(A)$ we denote the domain and the spectrum of an operator A, respectively.

Let S be a normal operator in L with the spectral resolution $E(s)$ defined on $\sigma(S)$, i.e.

$$S = \int_{\sigma(S)} s\,dE(s),$$

and let $W(s)$ be a function defined on $\sigma(S)$ whose values are linear operators in $\mathbf{C}^n$. Suppose that $W(s)$ is E-integrable in the Riemann sense on each finite interval. Consider the subset $D \subseteq H$ of elements $u \in H$ defined by (1.1) satisfying the conditions $(y_j, y_k) = 0$, $j \neq k$, and

$$\sum_j \int_{\sigma(S)} \|W(z)h_j\|_{\mathbf{C}^n}^2\, d(E(z)y_j, y_j)_L < \infty. \tag{1.3}$$

Define on D an operator $W(S)$ by the formula

$$W(S)u = \int_{\sigma(S)} W(s) \otimes dE(s)u. \tag{1.4}$$

This means that according to (1.1)

$$W(S)u = \sum_j \int_{\sigma(S)} W(s)h_j \otimes dE(s)y_j.$$

Besides, the integral is the limit in the norm $\|.\|_H$ of the consequence of the sums

$$\sum_m W(s_m)h_j \otimes E(\Delta_m)y_j,$$
$$\cup \Delta_m = \sigma(S), \; s_m \in \Delta_j$$

as the measure of $\max_m \Delta_m$ tends to zero.

Using relation (1.2) we can easily check that the equality $D = D(W(S))$ is true.

Our aim is to investigate operator-valued functions of the form (1.4). Below we will give a few relevant examples.

Let for instance

$$W(s) = \sum_{k=0}^{m} c_k s^k, \; (m < \infty)$$

where c_k are constant matrices. Then

$$W(S) = \sum_{k=0}^{m} c_k \otimes S^k.$$

By $I_H, I_L, I_{\mathbf{C}^n}$ we denote the identity operators in $H, L,$ and $\mathbf{C}^n$, respectively.

By the definition of $g(A)$ for a matrix A(see Section 1.1), the following relation can be written

$$g(W(s)) = \sqrt{N^2(W(s)) - \sum_{k=1}^{n} |\lambda_k(W(s))|^2}.$$

where $\lambda_k(W(s))$ are eigenvalues of $W(s)$. Besides, due to Corollary 1.3.7

$$g(A) \le \frac{N(A - A^*)}{\sqrt{2}} = \sqrt{2}N(A_I), \ A_I = \frac{1}{2i}(A - A^*) \qquad (1.5)$$

We will apply also Corollary 1.3.8 which implies

$$g(A + zI_{C^n}) = g(A) \qquad (1.6)$$

for any complex number z.

6.1.2 A norm estimation

For a normal operator S in L put

$$S_0 = S \otimes I_{C^n}.$$

It is clear that S_0 is a normal operator in H. We will say that λ is a regular point of $W(S) - S_0^{-\nu}\lambda$ if this operator is invertible.

Theorem 6.1.1 *Let $W(S)$ be a linear operator of the shape (1.4). Then for any $\nu \ge 0$ and a regular point λ of the operator $W(S) - S_0^{-\nu}\lambda$, the following inequality is true:*

$$\|(W(S) - S_0^{-\nu}\lambda)^{-1}\| \le d(\lambda, \nu),$$

with the notation

$$d(\lambda, \nu) \equiv \sup_{s \in \sigma(S)} \sum_{k=0}^{n-1} \frac{g^k(W(s))\gamma_{n,k}}{\rho^{k+1}(W(s), \lambda s^{-\nu})} \tag{1.7}$$

and $\rho(W(s), \lambda s^{-\nu})$ *is the distance between the spectrum* $\sigma(W(s))$ *of the matrix* $W(s)$ *and a point* $z = s^{-\nu}\lambda \in \mathbf{C}$.

Naturally, if $\nu \neq 0$, then it is assumed that operator S and S_0 have the bounded inverse ones.

The proof of this theorem is presented in Section 6.3.

6.2   Particular cases of Theorem 6.1.1

Let

$$W(s) = sI_{\mathbf{C}^n} + c,$$

where c is an $n \times n$-matrix , i e

$$W(S) = S \otimes I_{\mathbf{C}^n} + I_L \otimes c. \tag{2.1}$$

According to (1.6) we have the equality $g(W(s)) = g(c)$. The relation (1.5) implies:

$$g(W(s)) = g(c) \leq \frac{N(c - c^*)}{\sqrt{2}}. \tag{2.2}$$

Due to Theorem 6.1.1 the inequality

$$\|(W(S) - I_H \lambda)^{-1}\| \leq d(\lambda, 0)$$

is valid where

$$d(\lambda, 0) \equiv \sup_{s \in \sigma(S)} \sum_{k=0}^{n-1} \frac{g^k(W(s))\gamma_{n,k}}{\rho^{k+1}(W(s), \lambda)}.$$

This inequality and (2.2) give

$$\|(W(S) - I_H \lambda)^{-1}\| \leq \sum_{k=0}^{n-1} g^k(c) \frac{\gamma_{n,k}}{\rho^{k+1}(W(S), \lambda)}. \tag{2.3}$$

Here

$$\rho(W(S), \lambda) = \inf_{s \in \sigma(S)} \rho(W(s), \lambda)$$

is the distance between λ and the spectrum of $W(S)$.

Let now $W(S)$ be defined by

$$W(S) = S \otimes I_{\mathbf{C}^n} + c_0 \otimes I_L + c_1 \otimes S^{-1} + \ldots + c_{m-1} \otimes S^{-m+1} \qquad (2.4)$$

for a natural m . Here c_k are matrices.

Since S^{-1} and therefore S^{-k} ; $k = 2, \ldots, m - 1$ are bounded operators, omitting simple calculations we obtain the equality

$$D(W(S)) = D(S_0).$$

One can write down

$$W(s) = s \otimes I_{\mathbf{C}^n} + c_0 + c_1 s^{-1} + \ldots + c_{m-1} s^{-m+1}.$$

According to (1.6)

$$g(W(s)) = g(c_0 + \ldots + c_{m-1} s^{-m+1}).$$

Let S be a nonselfadjoint normal operator. Then thanks to the definition of $g(W(s))$ (see Subsection 6.1.1) we easily obtain the relation

$$v \equiv \sup_{s \in \sigma(S)} g(W(s)) \leq \sum_{k=0}^{m-1} N(c_k) \sup_{s \in \sigma(S)} |s|^{-k}.$$

Hence,

$$v \equiv \sup_{s \in \sigma(S)} g(W(s)) \leq v(S, W),$$

where

$$v(S, W) = \sum_{k=0}^{m-1} N(c_k) \|S^{-1}\|_L^k. \qquad (2.5)$$

Theorem 6.1.1. yields

$$\|(W(S) - S_0^{-m+1}\lambda)^{-1}\|_H \leq \sum_{k=0}^{n-1} \frac{v^k \gamma_{n,k}}{\delta^{k+1}(\lambda, W(S), m-1)} \qquad (2.6)$$

where

$$\delta(\lambda, W(S), m-1) \equiv \inf_{s \in \sigma(S)} \rho(W(s), \frac{\lambda}{s^{m-1}}).$$

That is

$$\delta(\lambda, W(S), m-1) \equiv \inf_{s \in \sigma(S)} \min_{k} |\lambda_k(W(s)) - \frac{\lambda}{s^{m-1}}|.$$

Let us suppose that S is a selfadjoint operator. Then its spectrum is real and thanks to (1.5):

$$\sup_{s \in \sigma(S)} g(W(s)) \leq v_1(S, W) \equiv \sqrt{1/2} \sum_{k=0}^{m-1} N(c_k^* - c_k) \|S^{-1}\|_L^k. \qquad (2.7)$$

Consequently, one can replace in (2.6) v by $v_1(S, W)$ if S is a selfadjoint operator.

6.3 Proof of Theorem 6.1.1

Lemma 6.3.1 *Let S be a normal operator in L with an orthogonal resolution of the identity $E(s)$, and $F(s)$ be a matrix-valued function defined on $\sigma(S)$. If $F(s)$ is E-integrable and*

$$m(F) \equiv \sup_{s \in \sigma(S)} \|F(s)\|_{\mathbf{C}^n} < \infty,$$

then the operator $F(S)$ defined in H by

$$F(S) = \int_{\sigma(S)} F(s) \otimes dE(s) \qquad (3.1)$$

is bounded and the estimate

$$\|F(S)\|_H \leq m(F)$$

is true.

Proof: We have

$$(F(S)u, F(S)u)_H = (\int_{\sigma(S)} F(s) \otimes dE(s)u, \int_{\sigma(S)} F(z) \otimes dE(z)u)_H$$

for any $u \in H$.

According to the definition of the scalar product on a tensor product of Hilbert spaces, for any u of the shape $u = y \otimes h$ with $y \in L$ and $h \in \mathbf{C}^n$, we can write down

$$(F(S)u, F(S)u)_H = \int_{\sigma(S)} \int_{\sigma(S)} (F^*(z)h, F(s)h)_{\mathbf{C}^n} (dE(s)y, dE(z)y)_L.$$

Take into account that

$$(dE(s)y, dE(z)y)_L = 0 \text{ if } s \neq z$$

and

$$(dE(s)y, dE(s)y)_L = (dE(s)y, y)_L \geq 0.$$

From this it follows

$$(F(S)u, F(S)u)_H \leq m^2(F)\|h\|_{\mathbf{C}^n}^2 \int_{\sigma(S)} (dE(s)y, y)_L.$$

Since

$$\int_{\sigma(S)} dE(s) = I_L,$$

due to the relationship (1.2) for the norms we obtain

$$\|F(S)(y \otimes h)\|_H \leq m(F)\|y \otimes h\|_H = m(F)\|u\|_H.$$

Extending this result according to (1.1), we get the stated result. $\square$

Corollary 6.3.2 *Let the matrix $W(s)$ have the inverse one $W^{-1}(s)$ for any $s \in \sigma(S)$, and let*

$$m_0 \equiv \sup_{s \in \sigma(S)} \|W^{-1}(s)\|_{\mathbf{C}^n} < \infty.$$

Then the operator $W(S)$ defined by (1.4) has the inverse one $W^{-1}(S)$. Moreover,

$$\|W^{-1}(S)\|_H \leq m_0$$

and

$$W^{-1}(S) = \int_{\sigma(S)} W^{-1}(s) \otimes dE(s).$$

In fact, set

$$K = \int_{\sigma(S)} W^{-1}(s) \otimes dE(s).$$

By Lemma 6.3.1 K is a bounded operator and $\|K\|_H \leq m_0$. Multiplying (1.4) from the left and from the right by K we can see that $K = W^{-1}(S)$. As claimed. $\Box$

Proof of Theorem 6.1.1: Employing Theorem 1.2.2, we have

$$\|(W(s) - s^{-\nu}\lambda I_{\mathbf{C}^n})^{-1}\|_{\mathbf{C}^n} \leq d(\lambda, \nu) \text{ for all } s \in \sigma(S).$$

Now Lemma 6.3.1 ensures the result. $\Box$

6.4 Perturbations of operators on tensor products

6.4.1 General operators

We will suppose that for a linear operator B acting in a Hilbert space H, the relation

$$q = \|W(S) - BS_0^{-\nu}\|_H < \infty \tag{4.1}$$

holds with some $\nu \geq 0$. Assume that

$$v \equiv \sup_{s \in \sigma(S)} g(W(s)) < \infty. \tag{4.2}$$

Theorem 6.4.1 *Let $W(S)$ be a linear operator defined by (1.4), and let a linear operator B in H satisfy (4.1). Put*

$$L = \{\mu \in \mathbf{C} : |\mu s^{-\nu} - \lambda_j(W(s))| \leq z(q, v) : s \in \sigma(S); j = 1, ..., n\},$$

where $\lambda_j(W(s))$ $(j = 1, ...n)$ are the eigenvalues of the matrix $W(s)$, and $z(q, v)$ is the extreme right-hand root of the algebraic equation

$$z^n = q \sum_{k=0}^{n-1} z^{n-k-1} \gamma_{n,k} v^k. \tag{4.3}$$

Then under (4.2), the spectrum of B lies in the set L.

In other words, under the hypothesis of Theorem 6.4.1 for any point $\mu \in \sigma(B)$ there is a point $s \in S$ and an eigenvalue $\lambda_j(W(s))$ of the matrix $W(s)$ such that $|\mu s^{-\nu} - \lambda_j(W(s))| \leq z(q, v)$.

The proof of Theorem 6.4.1 is given in Subsection 6.4.3.

If $\nu = 0$, then Theorem 6.4.1 implies

Corollary 6.4.2 *Let the conditions (4.2) and*

$$q = \|W(S) - B\| < \infty \tag{4.4}$$

be satisfied, then

$$sv_{W(S)}(B) \equiv \sup_{\mu \in \sigma(B)} \inf_{\lambda \in \sigma(W(S))} |\mu - \lambda| \leq z(q, v).$$

Let us suppose that $W(s) = sI_{\mathbf{C}^n}$, i.e. $W(S) = S \otimes I_{\mathbf{C}^n} = S_0$ is a normal operator, and let (4.4) hold. Take into account that $g(sI_{\mathbf{C}^n}) = 0$, and therefore $z(q, v) = q$. Hence due to Corollary 6.4.2 $sv_{S_0}(B) \leq q$. If both operators S_0 and B are normal, we obtain under (4.4) the bound $hd(S_0, B) \leq q$.

This means that Theorem 6.4.1 extends the well-known relationship for selfadjoint operators from (Kato, 1966) mentioned in Chapter 5 .

Again put

$$w(n) = \sum_{j=0}^{n-1} \gamma_{n,j}.$$

Setting $z = vy$ to (4.3), we have

$$y^n = \frac{q}{v} \sum_{k=0}^{n-1} y^{n-k-1} \gamma_{n,k}.$$

Let $qw(n) \leq v$. Then due to Lemma 4.3.2 the extreme right root y_0 of this equation satisfies the inequality

$$y_0 \leq \sqrt[n]{\frac{qw(n)}{v}}.$$

Since $z(q, v) = vy_0$, we obtain

$$z(q, v) \leq \sqrt[n]{qv^{n-1}w(n)}.$$

This result and Theorem 6.4.1 imply

Corollary 6.4.3 *Let $W(S)$ be a linear operator defined by (1.4), and let a linear operator B in H satisfy (4.1). Let additionally, (4.2) and the inequality $qw(n) \leq v$ hold. Then the spectrum of B lies in the set*

$$M = \{\mu \in \mathbf{C} : |\mu s^{-\nu} - \lambda_j(W(s))| \leq \sqrt[n]{qv^{n-1}w(n)}; \; s \in \sigma(S); j = 1, ..., n\}.$$

If $\nu = 0$, then the last result implies the following

Corollary 6.4.4 *Let the conditions (4.2), (4.4) and $qw(n) \leq v$ be satisfied, then*

$$sv_{W(S)}(B) \leq \sqrt[n]{qv^{n-1}w(n)}.$$

6.4.2 Operators with discrete spectra

Let us suppose that S has a discrete spectrum so that there are orthoprojectors P_k, $k = 1, 2, ...$ in L such that

$$P_j P_k = 0 \text{ if } k \neq j$$

and

$$S = \sum_{k=1}^{\infty} \lambda_k(S) P_k.$$

Besides we define $W(S)$ by

$$W(S)u = \sum_{k=1}^{\infty} (W_k \otimes P_k)u \text{ for all } u \in D(W(S)), \qquad (4.5)$$

where W_k, $k = 1, 2, ...$ are linear operators acting in $\mathbf{C}^n$ and $D(W(S))$ is defined as

$$D(W(S)) = \{u \in H : \sum_{k=1}^{\infty} \|(W_k \otimes P_k)u\|_H^2 < \infty\}.$$

This means that each u defined by (1.1) with $(y_j, y_k)_L = 0$ for $j \neq k$ belongs to $D(W(S))$ iff

$$\sum_{j} \sum_{k=1}^{\infty} \|W_k h_j\|_{\mathbf{C}^n}^2 \|P_k y_j\|_L^2 < \infty.$$

We assume that

$$v = \sup_{k=1,2,\ldots} g(W_k) < \infty. \tag{4.6}$$

The following result immediately follows from Theorem 6.4.1

Corollary 6.4.5   *Let a normal operator S in L have the discrete spectrum, and let $W(S)$ be defined by the series (4.5). Let (4.1) and (4.6) hold. Put*

$$L_k^d = \{\mu \in \mathbf{C} : |\mu \lambda_k^{-\nu}(S) - \lambda_j(W_k)| \le z(q,v) : j = 1,\ldots,n\} \tag{4.7}$$

where $z(q,v)$ is the extreme right zero of the algebraic equation (4.3) and $\lambda_j(W_k)$, $j = 1,\ldots,n$ are eigenvalues of the matrix W_k. Then the spectrum of the operator B lies in the set

$$\cup_{k=1}^{\infty} L_k^d.$$

Now we will exploit the estimation for the root $z(q,v)$ obtained in the previous subsection.

Corollary 6.4.6 *Let a normal operator S in L have the discrete spectrum, and let $W(S)$ be defined by the series (4.5). Let the relations (4.1), (4.6) and $qw(n) \le v$ hold. Put*

$$M_k^d = \{\mu \in \mathbf{C} : |\mu \lambda_k^{-\nu}(S) - \lambda_j(W_k)| \le \sqrt[n]{qv^{n-1}w(n)}, \ j = 1,\ldots,n\}$$

Then the spectrum of B lies in the set

$$\cup_{k=1}^{n} M_k^d.$$

If $\nu = 0$, then this result ensures the following

Corollary 6.4.7   *Let a normal operator S in L have the discrete spectrum, and let $W(S)$ be defined by the series (4.5). If the conditions (4.4), (4.6) and $qw(n) \le v$ are satisfied, then*

$$sv_{W(S)}(B) \le \sqrt[n]{qv^{n-1}w(n)}.$$

6.4.3   Proof of Theorem 6.4.1

Set

$$G_n(v,y) = \sum_{k=0}^{n-1} v^k y^{k+1} \gamma_{n,k}.$$

for a non-negative scalar variable y.

Lemma 6.4.8 *Under the hypothesis of Theorem 6.3.1 any point μ of the spectrum of the operator B satisfies the inequality*

$$1 \le qG_n(v, \frac{1}{\delta(\mu, W(S), \nu)}), \qquad (4.8)$$

where

$$\delta(\mu, W(S), \nu) = \inf_{s \in \sigma(S)} \rho(W(s), s^{-\nu}\mu).$$

Proof: One can rewrite the condition (4.1) in the form

$$q = \|(W(S)S_0^\nu - B)S_0^{-\nu}\|_H.$$

By Lemma 4.1.3 we can assert that each point $\mu \in \sigma(B)$ satisfies the inequality

$$q \, \| \, S_0^\nu R_\mu(A) \, \| \ge 1$$

with

$$A = W(S)S_0^\nu.$$

Thus,

$$q\|S_0^\nu (W(S)S_0^\nu - \mu I_H)^{-1}\|_H = q\|(W(S) - \mu S_0^{-\nu})^{-1}\|_H \ge 1. \qquad (4.9)$$

Theorem 6.1.1 gives:

$$\|(W(S) - S_0^{-\nu}\lambda)^{-1}\| \le d(\lambda, \nu).$$

Take into account the condition (4.2). Then

$$d(\lambda, \nu) \le \sum_{k=0}^{n-1} v^k \gamma_{n,k} \sup_{s \in \sigma(S)} \frac{1}{\rho^{k+1}(W(s), \lambda s^{-\nu})} =$$

$$= \sum_{k=0}^{n-1} v^k \gamma_{n,k} \frac{1}{\delta^{k+1}(\lambda, W(S), \nu)}.$$

We thus get

$$\|(W(S) - S_0^{-\nu}\lambda)^{-1}\| \leq \sum_{k=0}^{n-1} \frac{v^k \gamma_{n,k}}{\delta^{k+1}(\lambda, W(S), \nu)}.$$

Now (4.9) ensures the stated result.□

Proof of Theorem 6.4.1: Taking into account that $z(q,v)$ is the root of (4.3) and dividing (4.3) by z^n, we conclude that

$$1 = \sum_{k=0}^{n-1} v^k z^{-k-1}(q,v) \gamma_{n,k}.$$

Comparing this equality with (4.8), we have the inequality

$$z(q,v) \geq \delta(\mu, W(S), \nu).$$

This inequality is equivalent to the assertion.□

6.5   Particular cases of Theorem 6.4.1

Let

$$W(S) = S \otimes I_{\mathbf{C}^n} + I_L \otimes c, \tag{5.1}$$

where c is a constant matrix. That is, $W(S)$ has the form (2.1). Due to (2.2) $v = g(c)$. Let a linear operator B in H satisfy the condition

$$q = \|B - S \otimes I_{\mathbf{C}^n} + I_L \otimes c\|_H < \infty.$$

Then Corollary 6.4.2 implies:

$$sv_{W(S)}B \leq z(q,v). \tag{5.2}$$

Let now $W(S)$ be defined by (2.4). That is,

$$W(S) = S \otimes I_{\mathbf{C}^n} + c_0 \otimes I_L + c_1 \otimes S^{-1} + \ldots + c_{m-1} \otimes S^{-m+1}$$

for a natural m. Then the inequality

$$v = \sup_{s \in \sigma(S)} g(W(s)) \le v(S, W)$$

holds, where $v(S, W)$ is defined by (2.5). If S is selfadjoint, then (2.7) is true.

Let an operator B in H satisfy

$$q = \|W(S) - BS_0^{-m+1}\| < \infty, \tag{5.3}$$

then due to Theorem 6.4.1 the spectrum of B lies in the set

$$L = \{\mu \in \mathbf{C} : |\mu s^{-m+1} - \lambda_j(W(s))| \le z(q, v(S, W)) :$$

$$s \in \sigma(S), \; j = 1, ..., n\}.$$

Here $z(q, v(S, W))$ is the extreme right root of (4.3) with $v = v(W, S)$.

If the spectrum of S is discrete, then (2.4) has the shape (4.5) with

$$W_k = \lambda_k(S)I_{C^n} + c_0 + ... + c_{m-1}\lambda_k^{-m+1}(S).$$

In this case

$$v = max_k g(W_k) \le v(S, W).$$

Let, besides, an operator B satisfy (5.3), then due to Corollary 6.4.3, the spectrum of B lies in the set

$$\cup_{k=1}^{\infty} L_k^d,$$

where

$$L_k^d = \{\mu \in \mathbf{C} : |\mu \lambda_k^{-m+1}(S) - \lambda_j(W_k)| \le z(q, v(S, W)) : j = 1, 2, ..., n\} \tag{5.4}$$

6.6   The differentiation operator

6.6.1   Preliminaries

Let $\alpha = (\alpha_1, ..., \alpha_n)$ be a multi-index. Then D^α is a partial derivative of the form

$$D^\alpha = \frac{\partial^{\alpha_1 + ... + \alpha_n}}{\partial x_1^{\alpha_1} .... \partial x_n^{\alpha_n}}.$$

Let now Ω be an open set in a real Euclidean space $\mathbf{R}^l$. Then

$$H^l(\Omega, \mathbf{C}^n) \equiv W_2^l(\Omega, \mathbf{C}^n)$$

is the *Sobolev space* of vector-valued functions $f \in L^2(\Omega, \mathbf{C}^n)$ which have distributional derivatives of order $\leq l$ all square integrable, with the scalar product

$$(f, g)_{H^l(\Omega, \mathbf{C}^n)} \equiv \int_\Omega \sum_{j=0}^l (D^j f(x), D^j g(x))_{\mathbf{C}^n} d\Omega$$

and the norm

$$\|f\|_{H^l(\Omega, \mathbf{C}^n)} = \sqrt{(f, f)_{H^l(\Omega, \mathbf{C}^n)}}.$$

Further, $H_0^l(\Omega, \mathbf{C}^n)$ is the space of all functions from $H^l(\Omega, \mathbf{C}^n)$ vanishing on the boundary $\partial\Omega$ of the set Ω.

Imbedding the Sobolev space $H_0^1(\Omega, \mathbf{C}^n)$ into the space $L^2(\Omega, \mathbf{C}^n)$ is compact if Ω is a bounded set with a sufficiently smooth boundary.

If $n = 1$, then we will write $H_0^1(\Omega, C^1) = H_0^1(\Omega)$ and $H_0^l(\Omega, C^1) = H_0^l(\Omega)$.

For functions $f(x) \in H_0^l(\Omega)$ and $g(x) \in H_0^l(\Omega)$ the formulas integrating by parts

$$\int_\Omega \frac{\partial f(x)}{\partial x_i} g(x) d\Omega = - \int_\Omega \frac{\partial g(x)}{\partial x_i} f(x) d\Omega \; ; i = 1, ..., m$$

are valid.

For more details about Sobolev's spaces see for instance (Henry, 1981) and references given therein.

An operator A in a Hilbert space *is symmetric* if

$$(Ax, y) = (y, Ax)$$

An operator A in a Hilbert space *is symmetric* if

$$(Ax, y) = (y, Ax)$$

for every $x, y \in D(A)$. A symmetric operator A is called *semibounded* (from below) if for any $x \in D(A)$

$$(Ax, x) \geq a\, (x, x).$$

If A is semibounded, then the operator $A_1 = A + kI$ with a sufficiently large positive k is positively defined. If an operator A_{10} is the selfadjoint extension of A_1, then the operator $A_{10} - kI$ is the selfadjoint extension of A. Hence, without loss of generality we suppose that A is positively defined. The following *Friedrichs extension theorem* is very useful (Ahiezer and Glazman, 1981, p.384).

Any positive symmetric operator A whose domain is dense in H has at least one positive selfadjoint extension

For more details about selfadjoint extension see for instance (Ahiezer and Glazman, 1981).

6.6.2 First order operators

In $L^2(a, b)$ the operator P which has the domain $D(P)$ and associates with each function $\phi(t)$ the function

$$P\phi = i\frac{d\phi}{dt}$$

is called the *differentiation operator.*

A necessary condition for a function ϕ to belong to the domain $D(P)$ of P is that

(A) the function $\phi(t)$ must be absolutely continuous in every closed subinterval of the interval (a, b), and $\phi(t)$ and $\phi'(t)$ must belong to $L^2(a, b)$.

We shall consider separately the cases of: a finite interval, the semi-axis, and the whole axis.

In the first two cases, the domain $D(P)$ of the operator P will consist of functions which satisfy, in addition to the condition (A), certain boundary conditions.

1. **A finite interval** If the interval is finite, we may take it to be $(0, 2\pi)$, and the boundary condition takes the form

$$\phi(0) = \theta\phi(2\pi) \text{ with } |\theta| = 1. \tag{6.1}$$

The set of all functions satisfying (A) and (6.1) is clearly dense in $L^2(0, 2\pi)$. Further, P is a symmetric (unbounded) operator, since for any $\phi(t)$ and $\psi(t)$ from $D(P)$,

$$(P\phi, \psi) = i \int_0^{2\pi} \phi'(t)\overline{\psi(t)}dt =$$

$$i[\phi(2\pi)\overline{\psi}(2\pi) - \phi(0)\overline{\psi(0)}] + \int_0^{2\pi} \phi(t)\overline{i\psi'(t)}dt = (\phi, P\psi),$$

and so the integrated part vanishes:

$$(P\phi, \psi) = (\phi, P\psi).$$

Therefore due to the Friedrichs theorem the operator P has the selfadjoint extension. It is well-known (see for instance Ahiezer and Glazman (1981, p. 59) that *a function ϕ belongs to the domain $D(P_\theta)$ of the selfadjoint extension P_θ of the operator P if and only if it satisfies the conditions (A) and (6.1). This operator has no continuous spectrum.*

2. **The semi-axis** $(0, \infty)$ It is not hard to show that in this case the differentiation operator is symmetric if and only if the boundary condition takes the form

$$\phi(0) = 0$$

but it is irreducible. *That is, it has no selfadjoint extensions* (Ahiezer and Glazman, 1981, p.160).

3. **The whole axis** For every function ϕ which satisfies the condition (A), in the case of the whole axis the boundary conditions

$$\lim_{t \to -\infty} \phi(t) = \lim_{t \to \infty} \phi(t) = 0$$

are automatically satisfied. Therefore, the *domain $D(P)$ is defined by a single condition (A)*, and it is easily shown that *P is a selfadjoint operator. This operator has no eigenvalues* (Ahiezer and Glazman,1981, p.162).

6.6.3   A second order differential operator

Let A be defined by

$$A\phi(x) = -\phi''(x),\ 0 < x < l$$

whenever ϕ is a twice differentiable function on $[0, l]$ with the boundary condition

$$\phi(0) = \phi(l) = 0.$$

Considering this set of functions as subset of $L^2(0, l)$, we have

$$(A\phi, \psi) = -\int_0^l \phi''(t)\psi(t)dt = -\int_0^l \phi(t)\psi''(t)dt = (\phi, A\psi)$$

and

$$(A\phi, \phi) = -\int_0^l \phi''(t)\phi(t)dt = \int_0^l \phi'^2(t)dt \geq 0.$$

By the Friedrichs extension theorem, A can be extended to a selfadjoint operator A_0. In this case (Henry, 1981),

$$D(A_0) = \{\phi \in L^2(0, l) : A\phi \in L^2(0, l) = H_0^1(0, l) \cap H^2(0, l)\},$$

and the spectrum $\sigma(A_0)$ consists of the simple values

$$\lambda_n(A) = \frac{n^2\pi^2}{l^2}\ ,\ n = 1, 2, ...$$

with corresponding eigenfunctions

$$\phi_n(x) = sin\frac{n^2\pi^2}{l^2}.$$

More generally, if Ω is a bounded domain in R^n with a smooth boundary and Δ is the Laplacian operator, defined by the formula

$$A\phi = -\Delta\phi$$

on the set of sufficiently smooth functions with a compact support in Ω, then one can extend A to a selfadjoint positive definite operator A_0 in $L^2(\Omega)$ with

$$D(A_0) = H_0^1(\Omega) \cap H^2(\Omega).$$

6.7 A matrix differential operator on a segment

Our examples in this and in the next two sections only demonstrate the applications of Theorems 6.4.1 and by themselves do not claim to be essential.

Let us consider in $\mathbf{C}^n$ the problem

$$d^2u/dx^2 + c(x)u = \lambda u, \quad \text{for } 0 < x < 1 \tag{7.1}$$

with the boundary condition

$$u(0) = u(1) = 0. \tag{7.2}$$

Here $c(x)$ is a bounded measurable $n \times n$-matrix-valued function. Adapting this problem to Theorem 6.4.1, we rewrite $c(x)$ in the form $c(x) = c_0 + c_1(x)$ with a constant matrix c_0. For instance we can take

$$c_0 = \int_0^1 c(x)dx \text{ and } c_1(x) = c(x) - c_0.$$

The space $H = L^2([0,1], \mathbf{C}^n)$ is pertinent to our aim, because

$$L^2([0,1], \mathbf{C}^n) = L^2[0,1] \otimes \mathbf{C}^n.$$

We recall that $L^2([0,1], \mathbf{C}^n)$ is the Hilbert space of functions defined on $[0,1]$ with values in $\mathbf{C}^n$ and the scalar product

$$(h,g)_H = \int_0^1 (h(x), g(x))_{\mathbf{C}^n} dx,$$

and

$$L^2[0,1] \equiv L^2([0,1], \mathbf{C}^1)$$

is the space of scalar-valued functions. Set

$$D(S) = \{u \in L = L^2[0,1] : u'' \in L, u(0) = u(1) = 0\}$$

where the derivatives are understood in the distributed sense. That is

$$D(S) = H_0^1(0,1) \cap H^2(0,1)$$

and

$$D(S_0) = H_0^1([0,1], C^n) \cap H^2([0,1], C^n).$$

Also, put

$$Sy = y'', \; y \in D(S).$$

Thus, S is the Friedrichs extension of the corresponding differential operator defined on sufficiently smooth functions satisfying the boundary condition (7.2).

Put $D(B) = D(S_0)$ and define an operator B by

$$Bu = u'' + c(x)u, \; u \in D(S_0).$$

Now consider the matrix-valued function

$$W(s) = s \otimes I_{C^n} + c_0.$$

We can write down

$$W(S)u = u'' + c_0 u = (S \otimes I_{C^n} + I_L \otimes c_0)u. \tag{7.3}$$

Put

$$q = \sup_{x \in [0,1]} \|c_1(x)\|_{C^n}.$$

Hence

$$\|W(S) - B\|_H = q.$$

The relation (1.6) implies

$$g(W(s)) = g(s + c_0) = g(c_0).$$

Now, Corollary 6.4.2 entails the inequality

$$sv_{W(S)}(B) \leq z(q, g(c_0)),$$

where $z(q, g(c_0))$ is the extreme right-hand root of the algebraic equation
(4.3) with $v = g(c_0)$. The eigenvalues of the operator $W(S)$ defined by
(7.2) are

$$\lambda_{jk}(W(S)) = \lambda_k(S) + \lambda_j(c_0) = -\pi^2 k^2 + \lambda_j(c_0),$$

$$k = 1, 2, \ldots; \quad j = 1, \ldots, n.$$

Here $\lambda_j(c_0)$ are eigenvalues of c_0. Thus, *the eigenvalues of the problem
(7.1), (7.2) can be enumerated as $\lambda_{jk}(B)$ in such a way that*

$$|\lambda_{jk}(B) + \pi^2 k^2 - \lambda_j(c_0)| \leq z(q, g(c_0)).$$

If

$$qw(n) = q \sum_{j=0}^{n-1} \gamma_{n,j} \leq g(c_0), \tag{7.4}$$

then, as proved in Subsection 6.4.1,

$$z(q, v) \leq \sqrt[n]{q g^{n-1}(c_0) w(n)}.$$

Therefore under (7.4), the inequality

$$|\lambda_{jk}(B) + \pi^2 k^2 - \lambda_j(c_0)| \leq \sqrt[n]{q g^{n-1}(c_0) w(n)}$$

is valid for any eigenvalue $\lambda_{jk}(B)$ of the problem (7.1), (7.2).

In particular, if c_0 is a normal matrix, then $g(c_0) = 0$ and $z(q, g(c_0)) = q$.
I.e.

$$|\lambda_{jk}(B) + \pi^2 k^2 - \lambda_j(c_0)| \leq \sup_{x \in [0,1]} \|c(x) - c_0\|_{C^n} = q.$$

Let for instance $n = 1$. That is, $c(x)$ is a scalar-valued (nonreal, in general)
bounded function. Take

$$c_0 = \int_0^1 c(x) dx$$

and

$$q = \sup_x |c(x) - c_0|.$$

Then each eigenvalue $\lambda_k(B)$ of the problem (7.1), (7.2) subordinates the inequality

$$|\lambda_k(B) + \pi^2 k^2 - c_0| \leq q.$$

Thus the results of this section supplement the well-known asymptotical estimates for scalar second order differential operators (see e.g. Levitan and Sargsjan (1991)).

6.8 A periodic differential operator

Consider in $\mathbf{C}^n$ the problem

$$a(x)u' + b(x)u = \lambda u \text{ for } 0 < x < 1 \tag{8.1}$$

with the boundary condition

$$u(0) = u(1). \tag{8.2}$$

Here $a(x), b(x)$ are bounded measurable $n \times n$-matrix-valued functions and

$$det\, a(x) \neq 0 \text{ for any } x \in [0, 1].$$

Rewrite $a(x)$ and $b(x)$ in the form

$$a(x) = a_0 + a_1(x) \text{ and } b(x) = b_0 + b_1(x)$$

with constant matrices a_0 and b_0. For instance we can take

$$a_0 = \int_0^1 a(x)dx \text{ and } a_1(x) = a(x) - a_0,$$

and so on.

Take

$$H = L^2([0, 1], \mathbf{C}^n) \equiv L^2[0, 1] \otimes \mathbf{C}^n$$

and

$$D(S) = \{y \in L^2[0, 1] : y' \in L^2[0, 1];\ y(0) = y(1)\},$$

where the derivatives are understood in the distributed sense. That is, $D(S)$ is the set of functions y from $H^1(0,1)$ satisfying $y(0) = y(1)$. Besides,

$$D(S_0) = \{u \in L^2([0,1], \mathbf{C}^n) : u' \in L^2([0,1], \mathbf{C}^n);\ u(0) = u(1)\}$$

coincides with the set of functions from $H^1([0,1], C^n)$ satisfying (8.2).

Since the periodic differentiation operator is not invertible, we define S by the formula

$$Sy = y' + \pi i y, \text{ for } y \in D(S).$$

That is, S is the Friedrichs extension of the corresponding differential operator defined on sufficiently smooth functions satisfying the periodic boundary condition. This operator is invertible.

Also define on $D(S_0)$ the operator

$$Bu = a(x)u' + b(x)u \equiv a(x)S_0 u - a(x)i\pi u + b(x)u.$$

We take $D(B) = D(S_0)$.

It is clear that the eigenfunctions of S are $exp(2\pi k i x)$, and the eigenvalues of S are

$$\lambda_k(S) = i\pi(2k+1);\ k = 0, \pm 1, \ldots. \tag{8.3}$$

Take

$$W(s) = a_0 + s^{-1}(b_0 - a_0 \pi i).$$

I.e.

$$W(S) = a_0 \otimes I_L + (b_0 - a_0 \pi i) \otimes S^{-1}.$$

This implies

$$W(S) = \sum_{k=-\infty}^{\infty} W_k \otimes P_k,$$

where P_k are the eigenprojectors of S and

$$W_k \equiv a_0 + (b_0 - a_0 \pi i)\lambda_k^{-1}(S).$$

Take into account (8.3). Then

$$W_k = \frac{2k}{2k+1}a_0 + \frac{1}{i\pi(2k+1)}b_0. \tag{8.4}$$

We can write down

$$\|BS_0^{-1} - W(S)\|_H = \|(a(x) - a_0) \otimes I_L -$$

$$[(a(x) - a_0)\pi i - (b(x) - b_0)] \otimes S^{-1}\|_H \leq q,$$

where

$$q \equiv \sup_{0 \leq x \leq 1} [\|a_1(x)\|_{\mathbf{C}^n} + \|b_1(x) - \pi i a_1(x)\|_{\mathbf{C}^n}\|S^{-1}\|_L]. \qquad (8.5)$$

The relation (8.3) gives the equality

$$\|S^{-1}\|_L = \frac{1}{\pi}.$$

It is easily seen that eigenvalues of $W(S)$ are

$$\lambda_{jk}(W(S)) = \lambda_j(W_k); \quad k = 0, \pm 1, \dots \; ; j = 1, \dots, n.$$

So that (8.4) gives

$$\lambda_{jk}(W(S)) = \lambda_j(\frac{2k}{2k+1}a_0 + \frac{1}{i\pi(2k+1)}b_0).$$

Further, according to (1.5) and (8.4)

$$g(W_k) \leq v \equiv \sqrt{1/2}[N(a_0 - a_0^*) + \frac{1}{\pi}N(b_0 + b_0^*)] \qquad (8.6)$$

By Corollary 6.4.5 *the eigenvalues of the problem (8.1), (8.2) can be enumerated as* $\lambda_{jk}(B)$ *in such a way that the inequalities*

$$|\frac{\lambda_{jk}(B)}{i\pi(2k+1)} - \lambda_j(W_k)| \leq z(q,v),$$

$$k = 0, \pm 1, 2, \dots; \; j = 1, \dots, n$$

are valid.

Here q and v are defined by (8.5) and (8.6), respectively. Also, $z(q,v)$ is the extreme right-hand root of (4.3). If

$$qw(n) = q \sum_{j=0}^{n-1} \gamma_{n,j} \leq v, \qquad (8.7)$$

then according to Corollary 4.3.3,

$$z(q,v) \leq \sqrt[n]{qv^{n-1}w(n)}.$$

Thus, under (8.7) the inequality

$$\left|\frac{\lambda_{jk}(B)}{i\pi(2k+1)} - \lambda_j(W_k)\right| \leq \sqrt[n]{qv^{n-1}w(n)}$$

$$k = 0, \pm 1, 2, ...; \ j = 1, ..., n$$

is valid for any eigenvalue $\lambda_{jk}(B)$ of the problem (8.1), (8.2).

In many practical cases $a(x) \equiv I_{C^n}$ and (8.1) has the form

$$u' + b(x)u = \lambda u \text{ for } 0 < x < 1. \tag{8.8}$$

In this case $W(s) = I_{C^n} + s^{-1}(b_0 - \pi i)$. I.e.,

$$W(S) = I_H + (b_0 - I_{C^n}\pi i) \otimes S^{-1}.$$

This implies

$$W(S) = \sum_{k=-\infty}^{\infty} W_k \otimes P_k,$$

where

$$W_k \equiv I_{C^n} + (b_0 - I_{C^n}\pi i)\lambda_k^{-1}(S).$$

Taking into account (8.3) ,we obtain

$$W_k = \frac{2k}{2k+1}I_{C^n} + \frac{1}{i\pi(2k+1)}b_0.$$

We can write down

$$\|BS_0^{-1} - W(S)\|_H = \|(b(x) - b_0) \otimes S^{-1}\|_H \leq q_1,$$

where

$$q_1 \equiv \sup_{0 \leq x \leq 1} \|b_0 - b(x)\|_{C^n}\|S^{-1}\|_L \tag{8.9}$$

It is easily seen that in the considered case eigenvalues of $W(S)$ are

$$\lambda_{jk}(W(S)) = \frac{2k}{2k+1} + \frac{1}{i\pi(2k+1)}\lambda_j(b_0).$$

Hence

$$g(W_k) \le v_1 \equiv \frac{g(b_0)}{\pi} \text{ for } k = 0, \pm 1, \pm 2, \dots . \tag{8.10}$$

Notice that thanks to (1.5) $g(W_k) \le \frac{1}{\sqrt{2}\pi} N(b_0 + b_0^*)$. It is clear that by Corollary 6.4.2 and (8.3) the eigenvalues of the problem (8.8), (8.2) can be enumerated as $\lambda_{jk}(B)$ in such a way that the following relations

$$|\frac{\lambda_{jk}(B)}{i\pi(2k+1)} - \lambda_j(W_k)| \le z(q_1, v_1)$$

$$k = 0, \pm 1, \pm 2, \dots; \ j = 1, \dots, n.$$

are true. Here q_1 and v_1 are defined by (8.9) and (8.10), respectively, and $z(q_1, v_1)$ is the extreme right-hand root of (4.3) with $q = q_1$ and $v = v_1$. Consequently, we can write down

$$|\frac{\lambda_{jk}(B)}{i\pi(2k+1)} - \frac{2k}{2k+1} - \frac{1}{i\pi(2k+1)} \lambda_j(b_0)| \le z(q_1, v_1),$$

$$k = 0, \pm 1, \pm 2, \dots; \ j = 1, \dots, n.$$

If

$$q_1 w(n) = q_1 \sum_{j=0}^{n-1} \gamma_{n,j} \le v_1, \tag{8.11}$$

then by Lemma 4.3.2 the inequality

$$z(q_1, v_1) \le \sqrt[n]{q_1 v_1^{n-1} w(n)}$$

is true. Thus, under (8.11) the inequality

$$|\frac{\lambda_{jk}(B)}{i\pi(2k+1)} - \frac{2k}{2k+1} - \frac{1}{i\pi(2k+1)} \lambda_j(b_0)| \le \sqrt[n]{q_1 v_1^{n-1} w(n)}$$

$$k = 0, \pm 1, \pm 2, \dots; \ j = 1, \dots, n$$

is valid for any eigenvalue $\lambda_{jk}(B)$ of the problem (8.8), (8.2).

In particular, if b_0 is a normal matrix, then $g(b_0) = 0$, and $z_1(q_1, v_1) = q_1$. Therefore, one can write

$$|\frac{\lambda_{jk}(B)}{i\pi(2k+1)} - \frac{2k}{2k+1} - \frac{1}{i\pi(2k+1)} \lambda_j(b_0)| \le q_1,$$

$$k = 0, \pm 1, 2, ...; \ j = 1, ..., n$$

is valid for any eigenvalue $\lambda_{jk}(B)$ of the problem (8.8), (8.2).

In particular, if b_0 is a normal matrix, then $g(b_0) = 0$, and $z_1(q_1, v_1) = q_1$. Therefore, one can write

$$\left| \frac{\lambda_{jk}(B)}{i\pi(2k+1)} - \frac{2k}{2k+1} - \frac{1}{i\pi(2k+1)} \lambda_j(b_0) \right| \leq q_1,$$

$$k = 0, \pm 1, 2, ...; \ j = 1, ..., n.$$

For instance let $n = 1$. That is, $b(x)$ is a scalar-valued (nonreal, in general) bounded function. Take

$$b_0 = \int_0^1 b(x) dx$$

and

$$q_1 \equiv \frac{1}{\pi} \sup_{0 \leq x \leq 1} |b_0 - b(x)|.$$

Then each eigenvalue $\lambda_k(B)$ of the problem (8.8), (8.2) subordinates the inequality

$$\left| \frac{\lambda_k(B)}{i\pi(2k+1)} - \frac{2k}{2k+1} - \frac{1}{i\pi(2k+1)} b_0 \right| \leq q_1.$$

6.9 A high order matrix differential operator on the real axis

In this section $I = I_H$ is the identity operator in H. Consider in $H = L^2(\mathbf{R}^1, \mathbf{C}^n)$, where $\mathbf{R}^1$ is the real axis, the differential operator

$$B = \sum_{k=0}^{m} a_{m-k}(x) \frac{d^k}{dx^k}, \quad -\infty < x < \infty. \tag{9.1}$$

Here $a_k(x)$ are $n \times n$-matrix-valued functions measurable and bounded on $\mathbf{R}^1$. Also $a_0 = I_{\mathbf{C}^n}$.

It is clear that

$$L^2(R^1, \mathbf{C}^n) = L^2(R^1) \otimes \mathbf{C}^n$$

where $L^2(R^1) = L$ is the space of scalar-valued functions.

The domain of the operator (9.1) is defined as

$$D(B) = \{u \in H = L^2(R^1, C^n) : u^{(k)} \in H, \ k = 1, ..., m\},$$

where the derivatives are understood in the distributed sense. That is

$$D(B) = H^m(R^1, C^n).$$

Take

$$D(S) = \{y \in L = L^2(R^1) : y' \in L\}.$$

I.e. $D(S) = H^1(R^1)$. Since the differentiation operator is not invertible in $L^2(R^1)$, we define S by the formula

$$Su = u' - u \text{ for } u \in D(S).$$

That is, S is the Friedrichs extension of the corresponding differential operator defined on sufficiently smooth functions from $L^2(R^1)$. This operator is invertible.

We can write down

$$B = \sum_{k=0}^{m} a_{m-k}(x) \sum_{k=0}^{m} (S_0 + I_H)^k = \sum_{k=0}^{m} a_{m-k}(x) \sum_{j=0}^{k} C_k^{k-j} S_0^j,$$

where C_k^j are binomial coefficients. This implies

$$B = \sum_{j=0}^{m} b_{m-j}(x) S_0^j,$$

where $b_0(x) = I_{C^n}$ and

$$b_{m-j}(x) = \sum_{k=j}^{m} a_{m-k}(x) C_k^{k-j} \text{ for } j = 0, ..., m-1.$$

Take $b_k(x)$ in the form:

$$b_k(x) = c_k + h_k(x), \ k = 1, ..., m-1,$$

where c_k are constant $n \times n$-matrices and $h_k(x) = b_k(x) - c_k$. For example, if $a_k(x)$ are real matrices, then we can take

$$c_k = \frac{1}{2}(\sup_x b_k(x) + \inf_x b_k(x)).$$

Put

$$W(s) = sI_{C^n} + c_0 + c_1 s^{-1} + \ldots + c_m s^{-m+1}.$$

Obviously, $W(S)$ has the form (2.4) with $D(W(S)) = D(S_0)$. It is clear that the spectrum of S is

$$\sigma(S) = \{it - 1 : t \in R^1\}. \tag{9.2}$$

Thus ,

$$\|S^{-1}\|_L = \sup_{t \in R^1} \frac{1}{|it - 1|} = 1 \tag{9.3}$$

and

$$BS_0^{-m+1} - W(S) = \sum_{k=0}^{m}(b_k(x) - c_k)S_0^{-k}.$$

According to (9.3)

$$\|BS_0^{-m+1} - W(S)\|_H \leq$$

$$\sum_{k=0}^{m} \sup_{x \in R^1} \|b_k(x) - c_k\|_{C^n}\|S_0^{-k}\|_H \leq q,$$

where

$$q = \sum_{k=0}^{m} \sup_{-\infty \leq x \leq \infty} \|b_k(x) - c_k\|_{C^n}.$$

Due to (2.5) and (9.3)

$$\sup_{s \in \sigma S} g(W(s)) \leq v(S, W) = \sum_{k=0}^{m-1} N_2(c_k).$$

Thanks to Theorem 6.4.1 and (9.2) we can assert that the spectrum of B lies in the set

$$L = \{\mu \in \mathbf{C} : |\mu - (it - 1)^{m-1}\lambda_j(W(it - 1))| \leq$$

$$z(q, v(S, W))|it - 1|^{m-1} , -\infty < t < \infty, j = 1, \ldots, n\}.$$

Here $z(q, v(S, W))$ is the extreme right-hand root of (4.3) with $v = v(S, W)$.

In other words, for any $\mu \in \sigma(B)$ there are a real number t and an eigenvalue $\lambda_j(W(it - 1))$ of the matrix $W(it - 1)$ such that the inequality

$$|\mu - (it - 1)^{m-1}\lambda_j(W(it - 1))| \leq z(q, v(S, W))|it - 1|^{m-1}$$

is true.

If

$$qw(n) = q\sum_{j=0}^{n-1} \gamma_{n,j} \leq v(S, W), \qquad (9.4)$$

then, as proved in Subsection 6.4.1,

$$z(q, v(S, W)) \leq \sqrt[n]{qv^{n-1}(S, W)w(n)}$$

and spectrum of B lies in the set

$$M = \{\mu \in \mathbf{C} : |\mu - (it - 1)^{m-1}\lambda_j(W(it - 1))| \leq$$
$$\sqrt[n]{qv^{n-1}(S, W)w(n)}|it - 1|^{m-1}; -\infty < t < \infty, \ j = 1, \ldots, n\},$$

In other words, under (9.4) for any $\mu \subset \sigma(B)$ there are a real number t and an eigenvalue $\lambda_j(W(it - 1))$ of the matrix $W(it - 1)$ such that the inequality

$$|\mu - (it - 1)^{m-1}\lambda_j(W(it - 1))| \leq \sqrt[n]{qv^{n-1}(S, W) w(n)}|it - 1|^{m-1}$$

is true.

Note that in the scalar case we have $g(W(s)) = 0$ and we can put $z(q, v(S, W)) = q$.

Let us consider the second order system. That is, $m = 2$ and

$$B = \frac{d^2}{dx^2} + a_1(x)\frac{d}{dx} + a_2(x), \quad -\infty < x < \infty. \qquad (9.5)$$

Here $a_1(x), a_2(x)$ are $n \times n$-matrix-valued functions measurable and bounded on $\mathbf{R}^1$.

The domain of the operator (9.5) is defined as

$$D(B) = \{u \in H = L^2(R^1, \mathbf{C}^n) : u^{(k)} \in H, \ k = 1, 2\}$$

where the derivatives are understood in the distributed sense. That is, $D(B) = H^2(R^1, C^n)$.

Again take $D(S) = H^1(R^1)$ and define S by the formula $Su = u' - u$. For the operator (9.5) one can write down

$$B = (S_0 + I)^2 + a_1(x)(S_0 + I) + a_2(x) =$$

$$S_0^2 + 2S_0 + a_1(x)(S_0 + I) + a_2(x).$$

This implies

$$B = S_0^2 + b_1(x)S_0 + b_2(x),$$

where

$$b_1(x) = 2I_{C^n} + a_1(x)$$

and

$$b_2(x) = I_{C^n} + a_1(x) + a_2(x).$$

Take $b_k(x)$, $k = 1, 2$ in the form: $b_k(x) = c_k + h_k(x)$, $k = 1, 2$, where c_k are constant $n \times n$-matrices, and $h_k(x) = b_k(x) - c_k$. Further, put

$$W(s) = sI_{C^n} + c_1 + c_1 s^{-1}. \tag{9.6}$$

I.e. $W(S)$ has the form (2.4) with $m = 2$ and $D(W(S)) = D(S_0)$. Moreover,

$$BS_0^{-1} - W(S) = h_1(x) + h_2(x)S_0^{-1}.$$

So that according to (9.3)

$$\|BS_0^{-1} - W(S)\|_H \leq q_1,$$

where

$$q_1 = \sup_{-\infty \leq x \leq \infty} [\|b_1(x) - c_1\|_{C^n} + \|b_2(x) - c_2\|_{C^n}]. \tag{9.7}$$

Due to (1.6)

$$g(W(s)) = g(c_1 + c_2 s^{-1}).$$

Further, (1.5) implies

$$g(W(s)) \leq \sqrt{1/2}[N(c_1 - c_1^*) + N(s^{-1}c_2 - \bar{s}^{-1}c_2^*)].$$

Thus,

$$g(W(s)) \leq v_1(s) \text{ for each } s \in \sigma(S)$$

with the notation

$$v_1(s) = \sqrt{1/2}[N(c_1 - c_1^*) + 2N(c_2)|s|^{-1}].$$

Consequently,

$$\sup_{s \in \sigma(S)} g(W(s)) \leq v_1(S, W),$$

where according to (9.3)

$$v_1(S, W) = \sup_{s \in \sigma(S)} v(s) = \sqrt{1/2}\,[N(c_1 - c_1^*) + 2N(c_2)]. \qquad (9.8)$$

Thanks to Theorem 6.4.1 one can assert that *the spectrum of the operator B defined by (9.5) lies in the set*

$$L = \{\mu \in \mathbf{C} : |\mu\,(it - 1)^{-1} - \lambda_j(W(it - 1))| \leq$$
$$z(q_1, v_1(S, W)); \; -\infty < t < \infty, \; j = 1, \dots, n\}.$$

Here $z(q_1, v_1(S, W))$ is the extreme right-hand root of (4.3) with $v = v_1(S, W)$ and $q = q_1$. Recall that q_1 is defined by (9.7) and $v_1(S, W)$ is defined by (9.8).

Due to (9.6) one can rewrite this set in the form

$$L = \{\mu \in \mathbf{C} : |\mu - (it - 1)^2 - \lambda_j(c_1(it - 1) + c_2)| \leq z(q_1, v_1(S, W))|it - 1|;$$

$$-\infty < t < \infty, j = 1, \dots, n\}.$$

In other words, for any $\mu \in \sigma(B)$ there are a real number t and an eigenvalue $\lambda_j(c_1(it - 1) + c_2)$ of the variable matrix $(it - 1)c_1 + c_2$ such that the inequality

$$|\mu - (it - 1)^2 - \lambda_j(c_1(it - 1) + c_2)| \leq z(q_1, v_1(S, W))|it - 1|$$

is true.

If

$$q_1 w(n) = q_1 \sum_{j=0}^{n-1} \gamma_{n,j} \le v_1(S, W), \tag{9.9}$$

then, according to Corollary 4.3.3,

$$z(q_1, v_1(S, W)) \le \sqrt[n]{q_1 v_1^{n-1}(S, W) w(n)}.$$

Thus, under (9.9) the spectrum of the operator (9.5) lies in the set

$$M = \{\mu \in \mathbf{C} : |\mu - (it - 1)^2 - \lambda_j(c_1(it - 1) + c_2)|$$

$$\le \sqrt[n]{q_1 v_1^{n-1}(S, W) w(n)} |it - 1| \, ; -\infty < t < \infty, j = 1, \dots, n\}.$$

In other words, under (9.9) for any $\mu \in \sigma(B)$ there are a real number t and an eigenvalue $\lambda_j((it - 1)c_1 + c_2)$ of the variable matrix $c_1(it - 1) + c_2$ such that the inequality

$$|\mu - (it - 1)^2 - \lambda_j((it - 1)c_1 + c_2)| \le \sqrt[n]{q_1 v_1^{n-1}(S, W) w(n)} |it - 1|$$

is true.

For instance let $n = 1$. That is, $b_1(x)$ and $b_2(x)$ are scalar-valued (nonreal, in general) bounded functions. Then $g(W(s)) = 0$ and we can assert that the spectrum of B defined by (9.5) lies in the set

$$L = \{\mu \in \mathbf{C} : |\mu - (it - 1)^2 - c_1(it - 1) + c_2)| \le q_1 |it - 1| \, ; -\infty < t < \infty\},$$

where

$$q_1 = \sup_{-\infty \le x \le \infty} (|b_1(x) - c_1| + |b_2(x) - c_2|).$$

6.10   Notes

The contents of this chapter are based on (Gil', 1991a) and (Gil' and Shargorodskay, 1991).

As noted above Theorem 6.4.1 extends the well-known Theorem 6.4.10 from Kato (1966) on perturbation of the spectrum of selfadjoint operators.

The results presented in Sections 6.6-6.8 supplement the well-known estimations for the spectrum of scalar differential operators (see e.g. Levitan and Sargsjan (1991) and Kato (1966)).

Chapter 7

Stability and Boundedness of Ordinary Differential Systems

In this chapter stability and boundedness criteria for linear and nonlinear systems of ordinary differential equations are proposed. These criteria are based on the norm estimations for matrix-valued functions obtained above. In many cases they are formulated in terms of Hurwitzness of auxiliary matrices. This means that we can apply the well known stability criteria for linear stationary systems (for example, Hurwitz's criterion) to the investigation of nonstationary systems.

In Section 7.1 the notation and preliminary results are given.

Section 7.2 deals with general linear systems and linear systems with slowly varying coefficients.

In Section 7.3 integrally small perturbations of linear systems are considered.

Section 7.4 analyzes stability of nonlinear systems with nonstationary leading linear parts.

In Section 7.5 the "freezing" method for linear systems is extended to apply to nonlinear nonautonomous systems.

Section 7.6 analyzes boundedness of nonlinear systems with nonstationary leading linear parts.

7.1   Notation

Let **C** be a Euclidean space with the norm

$$\|x\| = \sqrt{|x_1|^2 + ... + |x_n|^2} \text{ for } x = (x_k) \in \mathbf{C}^n,$$

let I be the unit matrix, and A be an $n \times n$-matrix. $\lambda_k(A), k = 1, \ldots, n$ denote eigenvalues of A including their multiplicities.

An essential role hereafter is played by the quantity $g(A)$ defined in Section 1.1. As above

$$\alpha(A) \doteq \max_{k=1,\ldots,n} \Re\lambda_k(A).$$

Corollary 1.2.3 gives for any $n \times n$-matrix A the following inequality:

$$\|exp[At]\| \leq \Gamma(t, A) \text{ for all } t \geq 0, \tag{1.1}$$

where

$$\Gamma(t, A) \equiv exp[\alpha(A)t] \sum_{k=0}^{n-1} \frac{g^k(A)t^k}{(k!)^{3/2}}. \tag{1.2}$$

Consider in $\mathbf{C}^n$ the differential equation

$$\dot{x}(t) = f(t, x(t)) \text{ for all } t \geq 0 \ (\dot{x}(t) \equiv dx/dt), \tag{1.3}$$

where $f : [0, \infty) \times \mathbf{C}^n \to \mathbf{C}^n$. Throughout this chapter we shall assume that the function f is of such a nature that the equation (1.3) has a unique solution over $[0, \infty)$ corresponding to each initial condition $x(0) = x_0$ and that this solution depends continuously on x_0. Recall (Vidyasagar, 1993) that $x_1 \in \mathbf{C}^n$ is said to be an equilibrium point of the system (1.3) if

$$f(t, x_1) \equiv 0 \text{ for all } t \geq 0.$$

Throughout this chapter we shall assume that 0 is an equilibrium point of the system (1.3). This assumption does not result in any loss of generality because if x_1 is an equilibrium point of (1.3), then 0 is an equilibrium point of the system

$$\dot{z}(t) = f_1(t, z(t)),$$

where

$$f_1(t, z(t)) = f(t, z(t) + x_1).$$

Definition 7.1.1 *The equilibrium point 0 is said to be stable in the sense of Lyapunov if, for each $\epsilon > 0$, there exists $\delta > 0$ such that the condition*

$$\|x(0)\| \leq \delta$$

implies

$$\|x(t)\| \leq \epsilon \text{ for all } t \geq 0.$$

Definition 7.1.2 *The equilibrium point 0 is said to be asymptotically stable if it is stable in the sense of Lyapunov and there exists a number $\delta_1 > 0$ such that the condition*

$$\|x(0)\| \leq \delta_1$$

implies

$$\|x(t)\| \to 0, \text{ as } t \to \infty.$$

Also, the set $B(\delta_1)$ defined by

$$B(\delta_1) = \{x \in \mathbf{C}^n : \|x\| \leq \delta_1\}$$

is called a stability domain (a region of attraction) for the equilibrium point 0.

Definition 7.1.3 *The equilibrium point 0 is said to be globally asymptotically stable if*

$$\|x(t)\| \to 0, \text{ as } t \to \infty$$

for any solution x of (1.3) (regardless of what $x(0)$ is).

A polynomial $P(\lambda)$ is said to be a Hurwitz one if all its roots lie in the open left plane.

A matrix A is said to be a Hurwitz one if all its eigenvalues lie in the open left plane.

For more detail about stability see for instance Vidyasagar (1993).

If (1.3) is a linear equation, then the asymptotical stability is equivalent to the global asymptotical stability (see for instance Daleckii and Krein (1971)).

7.2 Linear systems

7.2.1 General systems

Consider in $\mathbf{C}^n$ the equation

$$dx/dt = A(t)x \text{ for } t \geq 0, \tag{2.1}$$

where $A(t)$ is a variable $n \times n$-matrix . Recall (Daleckii and Krein, 1971) that an operator $U(t,s) : \mathbf{C}^n \to \mathbf{C}^n$ is called the *evolution operator* of the equation (2.1) if

$$U(t,s)x(s) = x(t)$$

for any solution $x(t)$ of (2.1).

Assume that

$$q(t,s) \equiv \|A(t) - A(s)\|$$

is a non-negative scalar-valued piecewise continuous function for $t, s \geq 0$. Set

$$\theta = \sup_{t \geq 0} \Gamma(t, A(t)) =$$

$$\sup_{t \geq 0} exp[\alpha(A(t)t)] \sum_{k=0}^{n-1} \frac{g^k(A(t))t^k}{(k!)^{3/2}}.$$

Theorem 7.2.1 *Let the inequality*

$$\zeta \equiv \sup_{t \geq 0} \int_0^t \Gamma(t - s, A(t))q(t,s)ds < 1$$

hold. Then (2.1) is stable. Moreover, the estimate

$$\|x(t)\| \le (1-\zeta)^{-1}\theta\|x(s)\| \text{ for all } t > s \ge 0$$

is true for any solution $x(t)$ of (2.1).

The proofs of Theorem 7.2.1 and 7.2.2 are given in Subsection 7.2.3

7.2.2 Lipschitz's conditions

Consider the system (2.1) assuming that matrix $A(t)$ satisfies the conditions

$$v = \sup_{t\ge 0} g(A(t)) < \infty, \tag{2.2}$$

and

$$\|A(t) - A(s)\| \le q_0|t-s| \text{ for } t,s \ge 0, \tag{2.3}$$

where q_0 is a constant. Denote by $z(q_0,v)$ the extreme right-hand root of the algebraic equation

$$z^{n+1} = q_0 \sum_{k=0}^{n-1} \frac{v^k(k+1)}{\sqrt{k!}} z^{n-k-1}. \tag{2.4}$$

Obviously it is the unique positive and simple root. Put

$$W(n) = \sum_{k=0}^{n-1} \frac{k+1}{\sqrt{k!}}. \tag{2.5}$$

Setting $z = vy$ in (2.4) and applying Lemma 4.3.2, we can assert that the following estimates are true:

$$z(q_0,v) \le q_0 W(n) \text{ if } q_0 W(n) \ge v, \tag{2.6a}$$

and

$$z(q_0,v) \le v^{1-1/n}[q_0 W(n)]^{1/n} \text{ if } q_0 W(n) \le v. \tag{2.6b}$$

Theorem 7.2.2 *Let the conditions (2.2) and (2.3) hold. And let the matrix $A(t) + z(q_0,v)I$ be a Hurwitz one for all $t \ge 0$. Then (2.1) is stable.*

7.2.3   Proofs of Theorems 7.2.1 and 7.2.2

Proof of Theorem 7.2.1: Exploiting the inequality (1.1), we obtain

$$\|exp[A(\tau)t]\| \leq \Gamma(t, A(\tau)) \text{ for all } t, \; \tau \geq 0. \tag{2.7}$$

Equation (2.1) can be rewritten in the form

$$dx/dt - A(\tau)x = [A(t) - A(\tau)]x$$

for an arbitrary fixed $\tau \geq 0$. This equation is equivalent to the following:

$$x(t) = exp[A(\tau)t]x(0) + \int_0^t exp[A(\tau)(t - t_1)][A(t_1) - A(\tau)]x(t_1)dt_1. \tag{2.8}$$

From this according to (2.7)

$$\|x(t)\| \leq \Gamma(t, A(\tau))\|x(0)\| + \int_0^t \Gamma(t - t_1, A(\tau))q(\tau, t_1)\|x(t_1)\|dt_1.$$

If $\tau = t$, then the definitions of ζ and θ imply

$$\sup_{t \geq 0} \|x(t)\| \leq \theta \|x(0)\| + sup_{t_1 \geq 0}\|x(t_1)\|\zeta.$$

Thus, by the inequality $\zeta < 1$ we arrive at the result for $s = 0$. It is similarly proved for any $s \leq t$. $\square$

To prove Theorem 7.2.2 we set

$$\rho \equiv - \inf_{t \geq 0} \alpha(A(t))$$

and

$$\Gamma_0(t) = \exp[-\rho t] \sum_{k=0}^{n-1} \frac{t^k v^k}{(k!)^{3/2}}. \tag{2.9}$$

Assuming that $\rho > 0$, denote also

$$\chi \doteq \int_0^\infty \Gamma_0(t)dt = \sum_{k=0}^{n-1} \frac{v^k}{\sqrt{k!}\rho^{1+k}} \tag{2.10}$$

and

$$\zeta_0 \doteq q_0 \int_0^\infty t\Gamma_0(t)dt = \sum_{k=0}^{n-1} \frac{(k+1)v^k}{\sqrt{k!}\rho^{2+k}}.$$

It is simple to check that under (2.2) and (2.3) $\zeta \leq \zeta_0$.

Lemma 7.2.3 *Let the conditions (2.2), (2.3) and $\zeta_0 < 1$ be fulfilled. Then the evolution operator $U(t, s)$ of (2.1) satisfies the estimate*

$$\int_0^t \|U(t, w)\| dw \le (1 - \zeta_0)^{-1} \chi \text{ for all } t \ge 0$$

Moreover, equation (2.1) is asymptotically stable.

Proof: It is clear that under the hypothesis of the lemma

$$\Gamma(t, A(s)) \le \Gamma_0(t) \text{ for all } t, s \ge 0.$$

Thus, according to (2.7)

$$\|exp[A(s)t]\| \le \Gamma_0(t) \text{ for all } t, s \ge 0.$$

This relation and (2.8) imply for $\tau = t$ the following inequality:

$$y(t, w) \le \Gamma_0(t - w) + q_0 \int_0^t \Gamma_0(t - u)(t - u)y(u, w)du,$$

where

$$y(t, w) = \|x(t)\| / \|x(w)\|. \tag{2.11}$$

The integration with respect to w yields

$$\int_0^t y(t, w)dw \le \chi +$$

$$q_0 \int_0^t \int_w^t \Gamma_0(t - u)(t - u)y(u, w)dudw \tag{2.12}$$

because

$$\int_0^t \Gamma_0(t - w)dw \le \int_0^\infty \Gamma_0(s)ds = \chi.$$

Obviously

$$\int_0^t \int_w^t \Gamma_0(t - u)(t - u)y(u, w)dudw =$$

$$\int_0^t \int_0^u \Gamma_0(t - u)(t - u)y(u, w)dudw. \tag{2.13}$$

Further, it is easy to see that

$$\int_0^t \Gamma_0(t-u)(t-u)du \le \int_0^\infty \Gamma_0(t)tdt = \zeta_0/q.$$

Now, due to (2.12) and (2.13) we can write down

$$\int_0^t y(t,w)dw \le \chi + q_0 \max_{u \le t} \int_0^u y(u,w)dw \int_0^t \Gamma_0(t-u)(t-u)du$$

$$\le \chi + \zeta_0 \max_{u \le t} \int_0^t y(u,w)dw.$$

Thus, due to the inequality $\zeta_0 < 1$.

$$\max_{t \ge 0} \int_0^t y(t,w)dw \le (1-\zeta_0)^{-1}\chi.$$

This result proves the lemma since $x(t)$ is an arbitrary solution.□

Lemma 7.2.4 *Let the conditions (2.2), (2.3) and $\zeta_0 < 1$ hold. Then the evolution operator of the equation (2.1) subordinates the inequality*

$$\|U(t,s)\| \le \theta(1-\zeta_0)^{-1}$$

for all $t,s \ge 0$. Moreover, equation (2.1) is asymptotically stable.

Proof: Put

$$y_0 = \sup_{t,s \ge 0} y(t,s),$$

where $y(t,s)$ is defined by (2.11). Using (2.7) and (2.8), with $\tau = t$, we have

$$y_0 \le \Gamma(t,A(t)) + q_0 y_0 \int_0^t \Gamma_0(t-u)(t-u)du$$

because $\Gamma(t,A(t) \le \Gamma_0(t)$. Due to the inequalities $\Gamma(t,A(t)) \le \theta$ and

$$\int_0^t \Gamma_0(t-u)(t-u)du \le \zeta_0/q_0$$

we get the relation

$$y_0 \le \theta + y_0\zeta_0,$$

that ensures the stated estimate that entails the stability conditions in the sense of Lyapunov.

To obtain the asymptotical stability conditions it is sufficient to make in (2.1) the substitution $x(t) = x_1(t)e^{-t\epsilon}$ with a positive ϵ and to apply our reasoning to the perturbed system. For a sufficiently small ϵ the perturbed system will be stable in the sense of Lyapunov. This implies that (2.1) is asymptotically stable. The proof is complete.□

Proof of Theorem 7.2.2: Under the hypothesis of the theorem we have the inequality $\rho \geq z(q_0, v)$ where $z(q_0, v)$ is the extreme right-hand root of (2.4). Dividing (2.4) by z^{n+1}, we can write down

$$1 > q_0 \sum_{k=0}^{n-1} (k+1)v^k(k!)^{-1/2}\rho^{-k-2} \equiv \zeta_0.$$

Now, Lemma 7.2.4 entails the result.□

7.2.4   Remarks

Theorem 7.2.2 and the bounds (2.6) entail the following results. Let the conditions (2.2) and (2.3) hold. If $q_0 W(n) > v$ and the matrix $A(t) + q_0 W(n)I$ is a Hurwitz one for all $t > 0$, then (2.1) is stable. If under conditions (2.2), (2.3) we have $q_0 W(n) \leq v$, then for the stability of (2.1) it is sufficient that

$$A(t) + v^{1-1/n}[q_0 W(n)]^{1/n} I$$

is a Hurwitz matrix for all ≥ 0.

Note that any solution of (2.1) under (2.2) and (2.3) satisfies the inequality

$$\|x(t)\| \leq \|x(0)\| \, a \exp[(-\rho + z(q_0, v))t] \text{ for all } t \geq 0.$$

Here a is a constant. In fact, substituting the equality

$$x(t) = y(t)\exp(-tc)$$

with a real c into (3.1), we get the equation

$$dy/dt = (A(t) + Ic)y. \tag{2.14}$$

Let
$$c = \rho - z(q_0, v).$$

Then we can affirm that the matrix $A(t) + I(c + z(q_0, v))$ is a a Hurwitz one. Corollary 1.3.8 implies

$$g(A(t) + Ic) = g(A(t)).$$

Therefore, due to Theorem 7.2.2 the equation (2.14) is stable, i.e.

$$\|y(t)\| \le a\|y(0)\|, \text{ for all } t \ge 0.$$

This ensures the stated result since $x(t) = y(t)\exp(-tc)$.

7.2.5   Example

Let us consider the equation

$$\ddot{y} + p(t)\dot{y} + w(t)y = 0. \tag{2.15}$$

Here $p = p(t)$ and $w = w(t)$ are non-negative scalar-valued functions with the property

$$|p(t) - p(s)| + |w(t) - w(s)| \le q_0|t - s| \text{ for all } t, s \ge 0.$$

Put

$$a_{11}(t) = -p(t), a_{12}(t) = -w(t), a_{21}(t) = 1, a_{22}(t) = 0.$$

If $p^2 < 4w$, then

$$g(A) = [1 + w^2 + p^2 - |\lambda_1(A)|^2 - |\lambda_2(A)|^2]^{1/2}$$

$$= [(1 - w)^2 + p^2]^{1/2} \le 1 + w, \tag{2.16}$$

because

$$|\lambda_1(A)|^2 = |\lambda_2(A)|^2 = w$$

in this case. Here $A = A(t)$.

If $p^2 \geq 4w$, then

$$g(A) = [1 + w^2 + p^2 - |\lambda_1(A)|^2 - |\lambda_2(A)|^2]^{1/2} = 1 + w, \qquad (2.17)$$

since

$$|\lambda_1(A)|^2 + |\lambda_2(A)|^2 = p^2 - 2w$$

in this case (see also Example 1.3.3). Thus,

$$g(A(t)) \leq v \equiv 1 + \sup_{t \geq 0} w(t) \text{ for all } t \geq 0 \qquad (2.18)$$

Let us suppose that

$$\rho = -\sup_{t \geq 0} \alpha(A(t)) > 0 \text{ and } v < \infty.$$

If

$$\zeta_0 = q_0 \rho^{-1} + q_0 v \rho^{-2} < 1,$$

then the equation (2.15) is stable by Theorem 7.2.1 .

7.3 Integrally small perturbations of linear systems

7.3.1 Statement of the result

Let $U(t, s)$ be the evolution operator of the equation

$$dx/dt = A(t)x, \ t \geq 0, \qquad (3.1)$$

where $A(t) = (a_{jk}(t))$ is a variable $n \times n$-matrix, again. It is assumed that the following estimate for $U(t, s)$ is satisfied:

$$\|U(t, s)\| \leq \phi(t, s) \text{ for all } t, s \geq 0, \qquad (3.2)$$

where $\phi(t, s)$ is a known integrable non-negative function. We will consider the equation

$$dx/dt = A(t)x + B(t)x, \qquad (3.3)$$

where $B(t) = (b_{jk}(t))$ is a variable $n \times n$-matrix as a perturbation of equation (3.1) when the quantity

$$q(J) \equiv \sup\{\|J(t)\| : t \geq 0\} \tag{3.4}$$

is sufficiently small. Here and below

$$J(t) = \int_0^t B(s)ds.$$

Denote

$$m(t) = \|A(t)J(t) - J(t)(A(t) + B(t))\|.$$

Theorem 7.3.1 *Let the conditions (3.2) and*

$$\eta(J) \equiv q(J) + \sup_{t \geq 0} \int_0^t \phi(t,s)m(s)ds < 1 \tag{3.5}$$

be satisfied. Then for any solution $x(t)$ of (3.3) the estimate

$$\|x(t)\| \leq \|x(0)\|\,(1 - \eta(J))^{-1} \max_{t \geq 0} \phi(t,0) \tag{3.6}$$

is true.

Proof: We can rewrite (3.1) in the form

$$x(t) = U(t,0)x(0) + \int_0^t U(t,s)B(s)x(s)ds. \tag{3.7}$$

Evidently,

$$\int_0^t \frac{d}{ds}[U(t,s)J(s)x(s)]ds = J(t)x(t),$$

since $U(t,t)$ is the unit matrix. Exploiting the equality

$$dU(t,s)/ds = -U(t,s)A(s)$$

(see for instance (Daleckii and Krein, 1971)), we obtain

$$J(t)x(t) = \int_0^t U(t,s)[-A(s)J(s)x(s)+$$

$$B(s)x(s) + J(s)\dot{x}(s)]ds. \tag{3.8}$$

Substituting the right part of (3.3) instead of $\dot{x}(s)$ into (3.8), we arrive at the equality

$$\int_0^t U(t,s)B(s)x(s)ds = J(t)x(t) + \int_0^t U(t,s)[A(s)J(s)-$$

$$J(s)(A(s) + B(s))]x(s)ds.$$

Now (3.7) and the latter equality entail

$$x(t) = U(t,0)x(0) + J(t)x(t) + \int_0^t U(t,s)[A(s)J(s)-$$

$$J(s)(A(s) + B(s))]x(s)ds.$$

From this according to (3.2) and (3.4) it follows that

$$\|x(t)\| \leq \|x(0)\|\phi(t,0) + q(J)\|x(t)\| + \int_0^t \phi(t,s)m(s)\|x(s)\|ds.$$

This inequality implies

$$y(t,0) \leq \phi(t,0) + q(J)y(t,0) + \int_0^t \phi(t,s)m(s)y(s,0) \leq$$

$$\phi(t,0) + \eta(J)\sup_{t\geq 0} y(t,0),$$

where $y(t,s)$ is defined by (2.11). Now the condition (3.5) ensures the result.□

7.3.2 Integrally small perturbations of systems with constant matrices

Assume that $A(t) = A$ is a constant matrix and consider the equation

$$dx/dt = Ax + B(t)x. \tag{3.9}$$

Suppose that

$$q(J) < 1 \tag{3.10}$$

and

$$m_0 \equiv \sup_{t\geq 0} \|AJ(t) - J(t)(A + B(t))\| < \infty.$$

Denote by $z(m_0, q(J))$ the extreme right-hand root of the equation

$$z^n = m_0(1 - q(J))^{-1} \sum_{k=0}^{n-1} \frac{g^k(A)}{\sqrt{j!}} z^{n-k-1}. \qquad (3.11)$$

Clearly, $z(m_0, q(J))$ is a unique positive and simple root. Using Lemma 4.3.2, we can obtain for $z(m_0, q(J))$ bounds which are analogous to (2.6).

Theorem 7.3.2 *Let condition (3.10) hold. If the matrix $A + z(m_0, q(J))I$ is a Hurwitz one, then (3.9) is stable.*

Proof: In the case of a constant matrix the evolution operator $U(t, s)$ of the equation (3.1) is equal to $exp[A(t - s)]$. By (1.1) we get

$$\|U(t, s)\| \le \phi(t, s) = \Gamma(t - s, A).$$

It is easy to show that

$$\int_0^t \Gamma(t - s, A)ds \le \int_0^\infty \Gamma(s, A)ds = \sum_{k=0}^{n-1} \frac{g^k(A)}{\sqrt{k!}|\alpha(A)|^{k+1}}.$$

That is, in the case of a constant matrix

$$\eta(J) = q(J) + \sup_{t \ge 0} \int_0^t \Gamma(t - s, A)m(s)ds \le$$

$$q(J) + m_0 \sum_{k=0}^{n-1} \frac{g^k(A)}{\sqrt{k!}|\alpha(A)|^{k+1}}. \qquad (3.12)$$

Further, under the hypothesis of the lemma, the relation

$$|\alpha(A)| > z(m_0, q(J))$$

is true. Dividing the algebraic equation (3.11) by z^n and taking into account that $z(m_0, q(J))$ is its root, we obtain

$$1 > m_0(1 - q(J))^{-1} \sum_{k=0}^{n-1} \frac{g^k(A)}{\sqrt{j!}|\alpha(A)|^{k+1}}.$$

The result is

$$1 > q(J) + m_0 \sum_{k=0}^{n-1} \frac{g^k(A)}{\sqrt{j!}|\alpha(A)|^{k+1}}.$$

This inequality and (3.12) give $\eta(J) < 1$. Now Theorem 7.3.1 implies the stated result.□

7.3.3 Periodic systems

Let $A(t)$ be a T-periodic matrix. Take

$$A = T^{-1} \int_0^T A(t)dt$$

and $B(t) = A(t) - A$, then (3.1) can be rewritten in the shape (3.9) with

$$J(T) = \int_0^T B(t)dt = 0.$$

Thus,

$$q(J) = \sup_{t \geq 0} \|J(t)\| = \sup_{0 \leq t \leq T} \|J(t)\|,$$

and

$$m_0 = \sup_{0 \leq t \leq T} \|AJ(t) - J(t)(A + B(t))\|.$$

Now Theorem 7.3.2 gives us the stability condition for periodic systems.

7.3.4 Example

Consider the equation (3.9) under the condition (3.10) and $n = 2$, i.e.
$A = (a_{jk})$; $j, k = 1, 2$. Due to (1.2)

$$g(A) \leq \sqrt{1/2}|a_{12} - \bar{a}_{21}|.$$

Equation (3.11) has in the considered case the shape

$$z^2 = (1 - q(J))^{-1} m_0 (z + g(A)).$$

Consequently,

$$z(m_0, q(J)) = \frac{m_0}{2(1 - q(J))} + \sqrt{\frac{m_0^2}{4(1 - q(J))^2} + \frac{g(A)m_0}{1 - q(J)}}. \tag{3.13}$$

Now we can apply Theorem 7.3.2.

In particular, consider the equation

$$d^2y/dt^2 + (a + b(t))dy/dt + y = 0, \tag{3.14}$$

where $a = constant > 0$ and $b(t)$ is a real function with the property

$$\sup_{t \geq 0} |j(t)| < 1,$$

where $j(t) = \int_0^t b(s)ds$. It is simple to show that

$$q(J) = \sup_{t \geq 0} |j(t)|.$$

Thus, the condition (3.10) is satisfied. Set $a_{11} = -a$; $a_{12} = -1$; $a_{21} = 1$; $a_{22} = 0$; $b_{11}(t) = -b(t)$ and $b_{jk}(t) = 0$ for the other indexes.

Omitting simple calculations, we can write down

$$m_0 = \sup_{t \geq 0} \|AJ(t) - J(t)(A + B(t))\| =$$

$$\sup_{t \geq 0} |j(t)| \left(|b(t)|/2 + \sqrt{b^2(t)/4 + 1}\right).$$

Besides, due to (2.16) and (2.17) $g(A) \leq 2$. So we can set $g(A) = 2$ in (3.14). If

$$\alpha(A) + z(m_0, q(J)) < 0,$$

then (3.14) is stable by Theorem 7.3.2. Here

$$\alpha(A) = -a/2 + \sqrt{a^2/4 - 1}$$

if $a > 2$, and $\alpha(A) = -a/2$ if $a \leq 2$.

7.4   Stability of nonlinear systems with nonstationary leading parts

7.4.1   Statement of the result

This section is devoted to nonlinear ordinary differential systems with non-stationary leading (principle) parts. Stability and global stability conditions are obtained. The results obtained below are made more precise in the case of equations with stationary principle parts.

Let us consider in $\mathbf{C}^n$ the equation

$$dx/dt = A(t)x + F(x,t), \ t \geq 0, \tag{4.1}$$

where $A(t)$ is a variable $n \times n$-matrix and F maps $\Omega(r) \times [0,\infty)$ into $\mathbf{C}^n$ with the property

$$\|F(h.t)\| \leq \nu(t)\|h\| \text{ for all } h \in \Omega(r) \text{ and } \geq 0. \tag{4.2}$$

Here $\nu(t)$ is a non-negative integrable function and

$$\Omega(r) = \{h \in \mathbf{C}^n : \|h\| \leq r\}.$$

It is assumed that

$$q(t,s) \equiv \|A(t) - A(s)\|, t, s \geq 0$$

is a piecewise continuous function.

As above, it is assumed that a solution of the equation (4.1) exists for an initial vector $x(0) \in \Omega(r)$, it is unique, and continuously depends on initial conditions.

Recall that the function $\Gamma(t, A)$ is defined by (1.1), and

$$\theta \equiv \sup_{t \geq 0} \Gamma(t, A(t)).$$

Theorem 7.4.1 *Let the conditions (4.2) and*

$$\zeta(F) \equiv \sup_{t \geq 0} \int_0^t \Gamma(t - s, A(t))[q(t,s) + \nu(s)]ds < 1 \tag{4.3}$$

be fulfilled. Then the zero solution of (4.1) is asymptotically stable. An initial vector $x(0)$ satisfying the inequality

$$(1 - \zeta(F))^{-1} \theta\|x(0)\| \leq r, \tag{4.4}$$

belong to the region of attraction of the zero solution. Moreover, under (4.4) the estimation

$$\|x(t)\| \leq (1 - \zeta(F))^{-1}\theta\|x(0)\| \text{ for all } t \geq 0$$

is true.

In particular, let in (4.2) $\Omega(r) = \mathbf{C}^n$, i.e. $r = \infty$. Then under (4.2) and (4.3), the zero solution of (4.1) is globally asymptotically stable.

7.4.2   Lipschitz's conditions

Assume that the following conditions are satisfied:

$$\|A(t) - A(s)\| \le q_0|t - s| \text{ for all } t, s \ge 0 \tag{4.5}$$

and

$$\|F(h, t)\| \le \nu_0\|h\| \text{ for all } h \in \Omega(r), t \ge 0. \tag{4.6}$$

Here q_0 and ν_0 are non-negative constants.

Suppose, in addition, that

$$v = \sup_{t \ge 0} g(A(t)) < \infty \tag{4.7}$$

and

$$\rho \equiv -\sup_{t \ge 0} \alpha(A(t)) > 0.$$

Theorem 7.4.2 *Let the conditions (4.5-4.7) and*

$$\zeta_0(F) \equiv \sum_{j=0}^{n-1} \frac{v^j}{\sqrt{j!}} \left(\frac{q_0(j+1)}{\rho^{j+2}} + \frac{\nu_0}{\rho^{j+1}} \right) < 1 \tag{4.8}$$

hold. Then the zero solution of (4.1) is asymptotically stable. An initial vector $x(0)$ satisfying

$$(1 - \zeta_0(F))^{-1}\theta\|x(0)\| \le r, \tag{4.9}$$

belongs to the region of attraction. Moreover, under (4.9) the estimation

$$\|x(t)\| \le (1 - \zeta_0(F))^{-1}\theta\|x(0)\| \text{ for all } t \ge 0. \tag{4.10}$$

is true .

Denote

$$b_j = q_0 \frac{jv^{j-1}}{\sqrt{(j-1)!}} + \nu_0 \frac{v^j}{\sqrt{j!}} \text{ for } j = 1, ..., n-1,$$

and

$$b_0 = \nu_0, \quad b_n = q_0 \frac{v^{n-1}}{\sqrt{(n-1)!}}.$$

By $z_0(F)$ the extreme right-hand (unique positive and simple) zero of the algebraic equation

$$z^{n+1} = \sum_{j=0}^{n-1} b_j z^{n-j} \qquad (4.11)$$

is denoted.

Theorem 7.4.3 *Let the conditions (4.5-4.7) hold. If the matrix $A(t) + z_0(F)I$ is Hurwitzian for all $t \geq 0$, then the zero solution of (4.1) is asymptotically stable, and the inequality $\zeta_0(F) < 1$ is valid. Moreover, any vector $x(0)$ satisfying (4.9) belongs to the region of attraction of the zero solution and the estimate (4.10) under (4.9) is true.*

Corollary 7.4.4 *Let (4.5-4.7) hold with $\Omega(r) = \mathbf{C}^n$ (i.e. $r = \infty$), and let $A(t) + z_0(F)I$ be a Hurwitz matrix for all $t \geq 0$. Then the zero solution of (4.1) is globally asymptotically stable.*

7.4.3 Proofs of Theorems 7.4.1, 7.4.2, and 7.4.3

Proof of Theorem 7.4.1: As follows from (1.1),

$$\|exp[A(\tau)t]\| \leq \Gamma(t, A(\tau)). \qquad (4.12)$$

We rewrite the equation (4.1) in the form

$$dx/dt - A(\tau)x = [A(t) - A(\tau)]x + F(x,t),$$

regarding arbitrary $\tau \geq 0$ as fixed. This equation is equivalent to the following:

$$x(t) = exp[A(\tau)t]x(0) + \int_0^t exp[(A(\tau)(t-s)][(A(s) - A(\tau))x(s) + F(x,s)]ds.$$

Since the solutions continuously depend on the initial vector,

$$\|x(s)\| \leq r \text{ for } s \in [0,t],$$

for sufficiently small t. The latter inequality implies due to (4.2) and (4.12) the relation

$$\|x(t)\| \leq \Gamma(t, A(\tau))\|x(0)\| + \int_0^t \Gamma(t-s, A(\tau))[q(\tau,s) + \nu(s)]\|x(s)\|ds.$$

For $\tau = t$ according to the definitions of $\zeta(F)$ and θ we have

$$sup_{s \leq t} \|x(s)\| < \theta \|x(0)\| + sup_{s \leq t} \|x(s)\| \zeta(F).$$

Consequently,

$$sup_{s \leq t} \|x(s)\| \leq \theta \|x(0)\| (1 - \zeta(F))^{-1}$$

for a sufficiently small t. The condition (4.9) ensures this bound for all $t \geq 0$. As claimed. $\square$

Proof of Theorem 7.4.2: Rewrite (4.1) in the form

$$x(t) = U(t,0)x(0) + \int_0^t U(t,s)F(x,s)ds.$$

Set for sufficiently small t_0

$$c = \sup_{t \leq t_0} \|x(t)\|.$$

Take into account that solutions continuously depend on initial vectors. Then (4.6) gives

$$c \leq \|U(t,0)\| \|x(0)\| + \nu_0 c \int_0^t \|U(t,s)\| ds \ (t \leq t_0).$$

Lemmas 7.2.3 and 7.2.4 imply

$$c \leq \theta(1 - \zeta_0)^{-1} \|x(0)\| + \chi(1 - \zeta_0)^{-1} \nu_0 c, \qquad (4.13)$$

where ζ_0 and χ are defined in Subsection 7.2.3. It is clear that

$$\zeta_0(F) = \zeta_0 + \nu_0 \chi.$$

The condition $\zeta_0(F) < 1$ implies

$$1 - \zeta_0 > \nu_0 \chi,$$

and therefore

$$\chi(1 - \zeta_0)^{-1} \nu_0 < 1.$$

Now (4.13) gives the relation

$$c \leq \theta(1 - \zeta_0)^{-1} \|x(0)\| (1 - \chi(1 - \zeta_0)^{-1} \nu_0)^{-1} =$$

$$\theta\|x(0)\|(1-\nu_0\chi-\zeta_0)^{-1} = \theta\|x(0)\|(1-\zeta_0(F))^{-1}.$$

Due to (4.9) this bound can be extended from $[0, t_0)$ to the whole positive half-line. This bound yields the stability in the sense of Lyapunov. To obtain the asymptotical stability it is sufficient to apply the reasoning from Subsection 7.2.4.□

Lemma 7.4.5 *Let $z_0(F)$ be the extreme right-hand (unique positive and simple) zero of the algebraic equation (4.11). Then the inequality*

$$\alpha(A(t)) + z_0(F) < 0 \text{ for all } t \geq 0$$

implies the inequality $\zeta_0(F) < 1$.

Proof: The hypothesis of the lemma entail the inequality

$$\rho \equiv -\sup \alpha(A(t)) > z_0(F).$$

Dividing (4.11) by z^{n+1}, and taking into account that $z_0(F)$ is its root, we arrive at the inequality

$$1 = \sum_{j=0}^{n} b_j z_0(F)^{-j-1} > \sum_{j=0}^{n} b_j \rho^{-j-1}.$$

According to (4.8)

$$\zeta_0(F) = \sum_{j=0}^{n} b_j \rho^{-j-1}$$

Thus, we have the stated result.□

The assertion of Theorem 7.4.3 immediately follows from Theorem 7.4.2 and Lemma 7.4.4

7.4.4 Nonlinear systems with stationary linear parts

Let $A(t) \equiv A$ be a constant matrix and let (4.6) hold. Then it is simple to show that (4.11) turns out to the equation

$$z^n = \nu_0 \sum_{j=0}^{n-1} \frac{g^j(A)}{\sqrt{j!}} z^{n-j-1}. \tag{4.14}$$

Thus, $z_0(F)$ coincides with the extreme right-hand (unique positive and simple) zero $z_1(F)$ of the algebraic equation (4.14). Further, for $A(t) \equiv A$ one can write down

$$\zeta_0(F) = \nu_0 \sum_{j=0}^{n-1} \frac{g^j(A)}{|\alpha(A)|^{j+1}\sqrt{j!}},$$

and

$$\theta = \max_{t\geq 0} exp[\alpha(A)t] \sum_{j=0}^{n-1} g^j(A)t^j (j!)^{-1/2}.$$

Corollary 7.4.6 *Let $A(t) \equiv A$ be a constant matrix. Let also (4.6) hold. If the matrix $A + z_1(F)I$ is a Hurwitz one, then the zero solution of (4.1) is asymptotically stable. Moreover, $\zeta_0(F) < 1$ and any vector $x(0)$ satisfying the inequality (4.9) belongs to the region of attraction of the zero solution.*

In particular, let (4.6) and (4.7) hold with $\Omega(r) = \mathbf{C}^n$ (i.e. $r = \infty$), then the zero solution of (4.1) is globally stable if $A + z_1(F)I$ is a Hurwitz matrix.

7.4.5 Second order nonautonomous equation

Consider the equation

$$\ddot{y} + p(t)\dot{y} + w(t)y = f(t, y, \dot{y}), \qquad (4.15)$$

where $p(t)$ and $w(t)$ as in Subsection 7.2.5 are non-negative scalar-valued functions with the property

$$|p(t) - p(s)| + |w(t) - w(s)| \leq q_0|t - s| \text{ for all } t, s \geq 0.$$

It is assumed that the scalar-valued function f in (4.14) satisfies the condition

$$|f(t, y, z)|^2 \leq \nu_0^2 \left(|y|^2 + |z|^2\right) \qquad (4.16)$$

for

$$|y|^2 + |z|^2 \leq r^2 \text{ and } t, s \geq 0.$$

As shown in Subsection 7.2.5 ,

$$g((A(t)) \leq v \equiv 1 + \sup_{t \geq 0} w(t).$$

Let

$$\alpha(A(t)) \leq -\rho < 0 \text{ for all } t \geq 0$$

and

$$\zeta_0(F) = q_0 \rho^{-2} + \nu_0 \rho^{-1} + v(2q_0 \rho^{-3} + \nu_0 \rho^{-2}) < 1.$$

Then by Theorem 7.4.2 the zero solution of (4.15) is asymptotically stable. Each vector $x(0) = (y(0), y'(0))$ satisfying (4.9) belongs to the region of attraction of the zero solution.

Note that for the considered equation

$$\theta \leq max_{t \geq 0} exp[-t\rho] \, (1 + vt)$$

7.5 Stability of nonlinear systems without leading linear parts

7.5.1 Notation and statement of the result

Let us consider in C^n the equation

$$dx/dt = B(x(t), t)x \quad t \geq 0, \tag{5.1}$$

where

$$B(h, t) = (b_{jk}(h, t))$$

is an $n \times n$-matrix for any $h \in C^n$ and $t \geq 0$.

Let $\Omega(r)$ be a ball with the radius r and with the center in zero. For any $h = (h_k) \in \Omega(r)$ and $t \geq 0$ we construct the matrix $T(h, t) = (T_{jk}(h, t))$ with the elements

$$T_{jk}(h, t) \equiv T_{jk} = \partial b_{jk}(h, t)/\partial t + \sum_{m=1}^{n} F_m \partial b_{jk}(h, t)/\partial h_m.$$

Here

$$F_j = F_j(h,t) = b_{j1}(h,t)h_1 + \ldots + b_{jn}(h,t)h_n.$$

Let us suppose that

$$q(r) \equiv \sup\{\|T(h,t)\| : h \in \Omega(r), t \geq 0\} < \infty \tag{5.2}$$

and

$$v(r) \equiv \sup\{g(B(h,t)) : h \in \Omega(r), t \geq 0\} < \infty. \tag{5.3}$$

Set

$$\theta(r) = \sup\{\Gamma(t, B(h,t)) : h \in \Omega(r), t \geq 0\} \tag{5.4}$$

and denote by $z(r)$ the extreme right-hand (unique positive and simple) root of the algebraic equation

$$z^{n+1} = q(r) \sum_{j=0}^{n-1} \frac{v^j(r)(j+1)}{\sqrt{j!}} z^{n-j-1}. \tag{5.5}$$

Obviously $z(r)$ is a unique positive and simple root. Besides due to Lemma 4.3.2, for $z(r)$ the estimations of the type (2.6) are true.

Theorem 7.5.1 *Let for some positive $r \leq \infty$, the conditions (5.2) and (5.3) hold. If, in addition, the matrix $B(h,t) + Iz(r)$ is a Hurwitz one for all $h \in \Omega(r)$, and $t \geq 0$, then the zero solution of (5.1) is asymptotically stable. Besides, the following relationships hold:*

$$\rho(r) \equiv -\sup\{\alpha(B(h,t) : h \in \Omega(r), t \geq 0\} > 0$$

and

$$\zeta(r) \equiv q(r) \sum_{j=0}^{n-1} \frac{(j+1)v^j(r)}{\sqrt{j!}\rho^{j+2}(r)} < 1. \tag{5.6}$$

Moreover, any initial vector satisfying the condition

$$\|x(0)\|\frac{\theta(r)}{1-\zeta(r)} \leq r, \tag{5.7}$$

belongs to the region of attraction of the zero solution and for any solution $x(t)$ under (5.7), the estimate

$$\|x(t)\| \leq \|x(0)\| \frac{\theta(r)}{1-\zeta(r)} \quad \text{for all } t \geq 0 \tag{5.8}$$

is valid.

7.5.2 Proof of Theorem 7.5.1

Lemma 7.5.2 *If $x(t) : [0, \infty) \to \Omega(r)$ is a solution of (5.1), then*

$$\|dB(x(t), t)/dt\| \le q(r) \text{ for all } t \ge 0.$$

Proof: We have

$$dB(x(t), t)/dt = \partial B(x(t), t)/\partial t + \sum_{m=1}^{n} B_m(x(t), t)\dot{x}_m,$$

where

$$B_m(x, t) = \partial B(x, t)/\partial x_m,$$

and $x_m(t)$ are coordinates of $x(t)$. But according to (5.1)

$$\dot{x}_m(t) = \sum_{j=1}^{n} b_{mj}(x, t)x_j \equiv F_m(x, t).$$

Thus,

$$dB(x(t), t)/dt - T(x(t), t).$$

This gives the stated result.□

Proof of Theorem 7.5.1: We introduce the linear equation

$$dx/dt = B(h(t), t)x, \tag{5.9}$$

where

$$h(t) : [0, \infty) \to \Omega(r)$$

is a differentiable function. If $h(t) = x(t)$ is a solution of (5.1), then (5.9) and (5.1) coincide.

The continuous dependence of solutions on initial data implies that under the condition $\|x(0)\| < r$ there is t_0 such that

$$x(t) \in \Omega(r) \text{ for } 0 \le t \le t_0.$$

Now we use Theorem 7.2.2 to estimate a solution of (5.9) and therefore to estimate a solution of (5.1). If we put $A(t) = B(h(t), t)$ with some function $h(t)$, then (5.9) coincides with (2.1).

Let us suppose that $h(t) \in \Omega(r)$ for all $t \geq 0$ and

$$h(t) = x(t) \text{ for } 0 \leq t \leq t_0.$$

Then

$$v = \sup_{t \geq 0} g(A(t)) = \sup_{t \geq 0} g(B(h(t), t) \leq v(r),$$

Besides, thanks to Lemma 7.5.2

$$q_0 = \sup_{t \leq t_0} \|\dot{A}(t)\| = \sup_{t \leq t_0} \|dB(x(t), t)/dt\| \leq q(r).$$

Thus, for (2.1) with $A(t) = B(x(t), t)$ we have $\theta \leq \theta(r)$ and $\zeta_0 \leq \zeta(r)$. Repeating the reasoning of the proof of Lemma 5.4, we will show that the Hurwitzness of $B(x(t), t) + z(r)I$ implies the inequality $\zeta(r) < 1$. Consequently, according to Lemma 7.2.4 the bound

$$\|x(t)\| \leq \theta(r)(1 - \zeta(r))^{-1}\|x(0)\| \text{ for } t \leq t_0.$$

is true. But the condition (5.7) allows us to extend this bound to the whole half-line $t \geq 0$. As claimed.$\square$

7.5.3   A second-order scalar equation

Let us consider the equation

$$\ddot{y} + f(y, \dot{y}, t)\dot{y} + w(y, \dot{y}, t)y = 0, \qquad (5.10)$$

assuming that the scalar-valued functions $f = f(y, z, t)$ and $w = w(y, z, t)$ are positive and differentiable with respect to all their arguments if

$$y^2 + z^2 \leq r^2$$

for some fixed positive number $r \leq \infty$ and all $t \geq 0$.

   We put

$$b_{11} = -f, \; b_{12} = -w, \; b_{21} = 1 \text{ and } b_{22} = 0.$$

Due to (2.18) one can take

$$v(r) = 1 + \sup\{w(y, z, t) : y^2 + z^2 \leq r^2; \; t \geq 0\}.$$

Clearly, it is assumed that the supremum is finite. Further,

$$F_1 = b_{11}z + b_{12}y = -fz - wy, \quad F_2 = b_{21}z + b_{22}y = z,$$

and in the considered case the matrix

$$T(x,t) = (T_{jk}) = (T_{jk}(x,t)); \quad j,k = 1,2$$

with $x = (y,z)$, has the entries

$$T_{21} = T_{22} = 0, \quad T_{11} = -f'_t - f'_y z + (fy + wz)f'_z$$

and

$$T_{12} = -w'_t - w'_y z + (fz + wy)w'_z.$$

Besides, let

$$q(r) = \sup\{|T_{11}(x,t)| + |T_{12}(x,t)| : \|x\|^2 = y^2 + z^2 \leq r^2, t \geq 0\} < \infty.$$

Moreover,

$$z(r) = q(r)/2 + \sqrt{q^2(r)/4 + q(r)v(r)}.$$

If

$$\rho(r) \equiv \inf\{\Re[f(y,z,t)/2 - \sqrt{f^2(y,z,t)/4 - w(y,z,t)}] :$$

$$\|x\|^2 = y^2 + z^2 \leq r^2, \ t \geq 0\} > z(r),$$

then due to Theorem 7.5.1 the zero solution of (5.10) is asymptotically stable, the inequality

$$\zeta(r) = q(r)(\rho^{-1}(r) + v(r)\rho^{-2}(r)) < 1$$

is true, and any vector $x(0)$ satisfying (5.7) with

$$\theta(r) = \sup\{\Gamma(t, B(h,t)) : h \in \Omega, t \geq 0\} \leq \max_{t \geq 0} exp[-\rho(r)t] \, (1 + v(r)t)$$

belongs to the region of attraction. Moreover, the estimate (5.8) is true.

7.6 Boundedness of solutions of quasilinear systems

7.6.1 Statement of the result

Let us consider in $\mathbf{C}^n$ the equation

$$dx/dt = A(t)x + F(x,t), \ (t \geq 0) \tag{6.1}$$

where $A(t)$ is a variable $n \times n$-matrix and F maps $\Omega(r) \times [0, \infty)$ into $\mathbf{C}^n$ with the property

$$\|F(h.t)\| \leq \nu(t)\|h\| + l(t) \text{ for all } h \in \Omega(r) \text{ and } t \geq 0, \tag{6.2}$$

Here $l(t)$ and $\nu(t)$ are positive piecewise continuous non-negative functions, and

$$\Omega(r) = \{h \in \mathbf{C}^n : \|h\| \leq r\},$$

again. As above it is assumed that

$$q(t,s) \equiv \|A(t) - A(s)\|$$

is a non-negative piecewise continuous function defined for $t, s \geq 0$.

Recall that function $\Gamma(t, A)$ is defined by (1.1), and

$$\theta \equiv sup_{t \geq 0}\Gamma(t, A(t)).$$

Put

$$L(F) \equiv \sup_{t \geq 0} \int_0^t \Gamma(t - s, A(t))l(s)ds.$$

Theorem 7.6.1 *Let the conditions (6.2) and*

$$\zeta(F) \equiv \sup_{t \geq 0} \int_0^t \Gamma(t - s, A(t))[q(t, s) + \nu(s)]ds < 1 \tag{6.3}$$

be fulfilled. Let also for a $x_0 \in \mathbf{C}^n$, the inequality

$$\frac{\theta\|x_0\| + L(F)}{1 - \zeta(F)} \leq r \tag{6.4}$$

hold. *Then a solution $x(t)$ of equation (6.1) with the initial condition $x(0) = x_0$ is bounded. Moreover, the estimate*

$$\|x(t)\| \le \frac{\theta\|x(0)\| + L(F)}{1 - \zeta(F)} \text{ for all } t \ge 0$$

is true.

In particular, let in (6.2) $\Omega(r) = \mathbf{C}^n$, i.e. $r = \infty$. If under (6.2) the inequality (6.3) holds, then (6.4) is automatically satisfied and any solution of (6.1) is bounded.

Corollary 7.6.2 *Let the conditions (6.2) and (6.3) hold. Let also*

$$(1 - \zeta(F))^{-1} L(F) < r.$$

Then (6.1) possesses bounded solutions.

In fact, taking sufficiently small initial vectors satisfying (6.4) and employing Theorem 7.6.1, we arrive at the last assertion.

7.6.2   Lipschitz's conditions

Assume that

$$\|A(t) - A(s)\| \le q_0|t - s| \text{ for all } t, s \ge 0. \tag{6.5}$$

and

$$\|F(h.t)\| \le \nu_0\|h_0\| + l_0 \text{ for all } h \in \Omega(r), t \ge 0. \tag{6.6}$$

Here q_0, l_0, and ν_0 are non-negative constants. Suppose, in addition, as in Subsection 7.2.2 that

$$v = \sup_{t \ge 0} g(A(t)) < \infty \tag{6.7}$$

and

$$\rho \equiv -\sup_{t \ge 0} \alpha(A(t)) > 0.$$

Finally, put

$$L_0 \equiv l_0 \sum_{j=0}^{n-1} \frac{v^j}{\sqrt{j!}\rho^{j+1}}.$$

Theorem 7.6.3 *Let the conditions (6.5-6.7),*

$$\zeta_0(F) \equiv \sum_{j=0}^{n-1} \frac{v^j}{\sqrt{j!}} \left(\frac{q_0(j+1)}{\rho^{j+2}} + \frac{v_0}{\rho^{j+1}} \right) < 1 \qquad (6.8)$$

be fulfilled. Let for a $x_0 \in \mathbf{C}^n$, the inequality

$$(1 - \zeta_0(F))^{-1}(\theta \|x(0)\| + L_0) \leq r, \qquad (6.9)$$

hold. Then a solution $x(t)$ of equation (6.1) with the initial condition $x(0) = x_0$ is bounded. Moreover, the estimate

$$\|x(t)\| \leq (1 - \zeta_0(F))^{-1}(\theta \|x_0\| + L_0) \text{ for all } t \geq 0. \qquad (6.10)$$

is true.

Corollary 7.6.4 *Let under the conditions (6.5) - (6.8), the inequality*

$$(1 - \zeta_0(F))^{-1} L_0 < r,$$

hold. Then (6.1) possesses bounded solutions.

Recall that by $z_0(F)$ the extreme right-hand (unique positive and simple) zero of the algebraic equation (4.11) is denoted.

Theorem 7.6.5 *Let the conditions (6.5 - 6.7) hold. If the matrix $A(t) + z_0(F)I$ is Hurwitzian for all $t \geq 0$, then the inequality $\zeta_0(F) < 1$ is valid, and under (6.9) a solution $x(t)$ of (6.1) with $x(0) = x_0$ is bounded. Moreover, the estimate (6.10) is true.*

Corollary 7.6.6 *Let (6.5)-(6.7) hold with $\Omega(r) = \mathbf{C}^n$ (i.e. $r = \infty$), and let $A(t) + z_0(F)I$ be a Hurwitz matrix for all $t \geq 0$. Then any solution of (6.1) is bounded and the estimate (6.10) is true.*

7.6.3   Proofs

Proof of Theorem 7.6.1: As follows from (1.1),

$$\|exp[A(\tau)t]\| \leq \Gamma(t, A(\tau)). \qquad (6.11)$$

We rewrite (6.1) in the form

$$dx/dt - A(\tau)x = [A(t) - A(\tau)]x + F(x,t)$$

for an arbitrary fixed $\tau \geq 0$. This equation is equivalent to the following:

$$x(t) = exp[A(\tau)t]x(0)+$$

$$\int_0^t exp[A(\tau)(t-s)][(A(s) - A(\tau))x(s) + F(x,s)]ds.$$

Due to (6.2) and (6.11)

$$\|x(t)\| \leq \Gamma(t, A(\tau))\|x(0)\| + \int_0^t \Gamma(t-s, A(\tau))) + \nu(s)]\|x(s)\|ds$$

$$+ \int_0^t \Gamma(t-s, A(\tau))l(s)ds, \ t \leq t_0$$

if

$$\|x(t)\| \leq r \text{ for } t \in [0, t_0].$$

For $\tau = t$ and according to the definitions of $L(F)$, $\zeta(F)$ and θ we have

$$sup_{t \leq t_0}\|x(t)\| < \theta\|x(0)\| + sup_{s \leq t_0}\|x(s)\|\zeta(F) + L(F)$$

if t_0 is sufficiently small, since solutions continuously depend on initial vectors. Consequently,

$$sup_{s \leq t}\|x(s)\| \leq \frac{\theta\|x(0)\| + L(F)}{1 - \zeta(F)}$$

for a sufficiently small t. The condition (6.9) ensures this bound for all $t \geq 0$. As claimed.$\square$

Proof of Theorem 7.6.3: Rewrite (6.1) in the form

$$x(t) = U(t,0)x(0) + \int_0^t U(t,s)F(x,s)ds,$$

where $U(t,s)$ is the evolution operator of the linear part of the equation. Set

$$c(t_0) = \sup_{0 \leq t \leq t_0} \|x(t)\|.$$

for a sufficiently small t_0.

Taking into account (6.6), we get

$$c(t_0) \le \|U(t,0)\|\|x(0)\| + (\nu_0 c(t_0) + l_0) \int_0^t \|U(t,s)\| ds$$

Employing Lemmas 7.2.3 and 7.2.4, we obtain

$$c(t_0) \le \theta(1 - \zeta_0)^{-1}\|x(0)\| + \chi(1 - \zeta_0)^{-1}(\nu_0 c(t_0) + l_0) \qquad (6.12)$$

for sufficiently small t_0.

Recall that ζ_0 and χ are defined in Subsection 7.2.2. It is clear that

$$\zeta_0(F) = \zeta_0 + \nu_0 \chi.$$

The condition $\zeta_0(F) < 1$ implies

$$1 - \zeta_0 > \nu_0 \chi,$$

and therefore

$$\chi(1 - \zeta_0)^{-1}\nu_0 < 1.$$

Consequently, (6.12) gives:

$$c(t_0) \le \frac{\theta\|x(0)\| + l_0 \chi}{1 - \zeta_0}\left(1 - \frac{\chi\nu_0}{1 - \zeta_0}\right)^{-1}.$$

That is,

$$c(t_0) \le \frac{\theta\|x(0)\| + l_0 \chi}{1 - \nu_0 \chi - \zeta_0}.$$

According to (6.9) we can extend this inequality to $[0, \infty)$. This inequality proves the stated result because $L_0 = \chi l_0$. $\square$

The assertion of Theorem 7.6.5 follows immediately follows from Theorem 7.6.3 and Theorem 7.4.4.

7.6.4   Boundedness of systems with stationary linear parts

Let $A(t) \equiv A$ be a constant matrix. Then it is simple to show that (4.11) becomes the algebraic equation (4.14). Thus, $z_0(F)$ coincides with the extreme right-hand (unique positive and simple) zero $z_1(F)$ of (4.14). Further, for $A(t) \equiv A$ we can write down

$$\zeta_0(F) = \nu_0 \sum_{j=0}^{n-1} \frac{g^j(A)}{|\alpha(A)|^{j+1}\sqrt{j!}},$$

$$L_0 = l_0 \sum_{j=0}^{n-1} \frac{g^j(A)}{|\alpha(A)|^{j+1}\sqrt{j!}}$$

and

$$\theta = \max_{t \geq 0} exp[\alpha(A)t] \sum_{j=0}^{n-1} \frac{g^j(A)t^j}{\sqrt{j!}}.$$

Exploiting Theorem 7.6.3 and omitting simple calculations we obtain the following two results:

Corollary 7.6.7 *Let $A(t) \equiv A$ be a constant matrix. Let also (6.6) hold. If the matrix $A + z_1(F)I$ is a Hurwitz one, then $\zeta_0(F) < 1$, and under (6.9), a solution $x(t)$ of (6.1) with $x(0) = x_0$ is bounded. Moreover, the estimate (6.10) is true.*

Corollary 7.6.8 *Let $A(t) \equiv A$ be a constant matrix. Let also (6.6) hold. If matrix $A + z_1(F)I$ is a Hurwitz one, then $\zeta_0(F) < 1$. If besides,*

$$(1 - \zeta_0(F))^{-1}L_0 < r,$$

then (6.1) possesses bounded solutions.

In particular, let (6.6) hold with $\Omega(r) = \mathbf{C}^n$ (i.e. $r = \infty$), then all solutions of (6.1) are bounded if $A + z_1(F)I$ is a Hurwitz matrix.

7.6.5 Example

Consider the equation

$$\ddot{y} + p(t)\dot{y} + w(t)y = f(t, y, \dot{y})$$

where $p(t)$ and $w(t)$ as in Subsection 7.2.5 are non-negative scalar-valued functions with the property

$$|p(t) - p(s)| + |w(t) - w(s)| \leq q_0|t - s| \text{ for all } t, s \geq 0.$$

Besides, it is assumed that the scalar-valued function f satisfies the condition

$$|f(t, y, z)|^2 \leq \nu_0^2(|y|^2 + |z|^2) + l_0 \text{ for all } t, s \geq 0$$

and

$$|y|^2 + |z|^2 \leq r^2.$$

As shown at Subsection 7.2.5, if

$$\sup_{t \geq 0} w(t) < \infty,$$

then

$$g((A(t)) \leq v \equiv 1 + \sup_{t \geq 0} w(t).$$

Assume that

$$\alpha(A(t)) \leq -\rho < 0 \text{ for all } t \geq 0.$$

Let

$$\zeta_0(F) = q_0\rho^{-2} + \nu_0\rho^{-1} + v(2q_0\rho^{-3} + \nu_0\rho^{-2}) < 1,$$

and inequality

$$\frac{L_0}{1 - \zeta_0(F)} < r,$$

hold. Then due to Theorem 7.6.5 the considered equation possesses bounded solutions.

Note that in the considered example

$$L_0 = l_0(\frac{1}{\rho} + \frac{v}{\rho^2}).$$

7.7   Notes

The contents of Section 7.2 are adapted from Gil' (1987a), Gil' (1990). Theorem 7.2.2 makes the well-known stability criterion by Vinograd (1983) somewhat more precise.

The results of Section 7.3 are new. As shown above, Theorems 7.2.2 and 7.3.2 allow us to apply Hurwitz's criterion to the investigation of non-stationary linear systems.

The material in Section 7.4 and Section 7.5 is taken from Gil' (1990), Gil' (1989a), and Gil'(1989b). Theorems 7.4.3 and 7.5.1 allow us to apply Hurwitz's criterion to nonlinear systems. Theorems 7.4.1, 7.4.2, and 7.5.1 give the bounds for the region of attraction. Besides, Corollary 7.4.6 supplements the absolute stability criterion from (Krasnoselskii and Pokrovskii, 1977).

The results presented in these sections are already applied to the predator-prey system (Gil', 1984c), oscillations of autogenerators (Gil' and Shargorodskay, 1986b) and limiting cycle (Gil' and Shargorodskay, 1986a).

The classical absolute stability criteria for other classes of nonlinearities can be found in Popov (1973), Naredra and Taylor (1973).

The contents of Section 7.6 are new.

Chapter 8

Stability of Retarded Systems

This chapter is devoted to the stability of systems of differential equations with delay.

In Section 8.1 some well-known notions of the theory of retarded equations are listed.

In Section 8.2 the main results of the chapter are presented.

Section 8.3 is devoted to absolute stability of equations with delay.

Proofs of assertions presented in Sections 8.2 and 8.3 are collected in Section 8.4.

In Section 8.5 an asymptotical behavior of solutions of nonlinear equations with stationary linear parts is examined.

8.1 Notation

Let us consider in a Euclidean space $\mathbf{C}^n$ with the norm $\|.\|_{\mathbf{C}^n}$ the equation

$$dx/dt = \int_0^\eta dR(\tau)x(t-\tau) + F(t, x_t) \text{ for all } t \geq 0 \qquad (1.1)$$

with a positive constant $\eta < \infty$. The symbol x_t means the vector-valued function $x(t + \tau)$ $(-\eta \leq \tau \leq 0)$ cf. (Hale, 1977), $R(\tau)$ is a matrix-valued function having a bounded variation, and the function F maps $[0, \infty) \times L^2([-\eta, \infty), \mathbf{C}^n)$ into $\mathbf{C}^n$.

Take the initial condition

$$x(t) = \Phi(t) \text{ for } -\eta \leq t < 0, \text{ and } x(+0) = \Phi(0), \qquad (1.2)$$

where $\Phi(t)$ is a given continuous vector-valued function. Put

$$\|x\|_L \equiv [\int_0^\infty \|x(t)\|_{\mathbf{C}^n}^2 \, dt]^{1/2}.$$

As usual $L^2([0, \infty), \mathbf{C}^n)$ is the space of n-dimensional vector-valued functions defined on the half-line $[0, \infty)$ with the norm $\|.\|_L$. Similarly, the space $L^2([-\eta, 0], \mathbf{C}^n)$ is defined.

*A solution of (1.1) on $[0, \infty)$ with initial value Φ is an absolutely con*tinuous function which satisfies this equation on $[0, \infty)$ almost everywhere over $[0, \infty)$, and (1.2) holds (see for instance Hale (1977), Myshkis (1972)).

We assume the solution existence and the uniqueness, although a priori estimates obtained below make it possible to prove the existence by the traditional scheme with the help of the Leray-Schauder principle.

Definition 8.1.1 *Suppose $F(0, t) = 0$ for $t \geq 0$. The solution $x = 0$ of the equation (1.1) is said to be stable if for any $\epsilon > 0$ there is a $\delta > 0$ such that the condition*

$$\|\Phi(t)\|_{\mathbf{C}^n} \leq \delta \text{ for all } -\eta \leq t \leq 0$$

implies the inequality

$$\|x(t)\|_{\mathbf{C}^n} \leq \epsilon \text{ for all } t \geq 0$$

and any solution $x(t)$ of (1.1) with the initial condition (1.2).

The solution $x = 0$ of the equation (1.1) is said to be asymptotically stable if it is stable and there is a constant $a_0 > 0$ such that the condition

$$\|\Phi(t)\|_{\mathbf{C}^n} \leq a_0 \text{ for all } -\eta \leq t \leq 0$$

implies the relationship

$$\lim_{t\to\infty} \|x(t)\|_{C^n} = 0$$

for any solution $x(t)$ of (1.1) with the initial condition (1.2).

Besides, the set

$$\{h \in C^n : \|h\|_{C^n} \leq a_0\}$$

is called a region of attraction (a stability domain) of the zero solution.

The solution $x = 0$ of the equation (1.1) is said to be globally asymptotically stable if it is asymptotically stable and its region of attraction coincides with C^n (i.e. $a_0 = \infty$).

For a positive number r we denote by $\Omega(r)$ *the set of bounded functions* h defined on $[-\eta, \infty)$ with values from C^n and satisfying $\|h(t)\|_{C^n} \leq r$ for all $t \geq -\eta$. That is,

$$\Omega(r) = \{h \in L^2([-\eta, \infty), C^n) : \|h(t)\|_{C^n} \leq r \text{ for all } t \geq -\eta\}.$$

Let us suppose that for some positive $r \leq \omega$ and $q = const < \infty$, the inequality

$$\|F(\cdot, u_t)\|_L \equiv [\int_0^\infty \|F(t, u_t)\|_{C^n}^2 \, dt]^{1/2} \leq q \, (\|u\|_L + \zeta(u))$$

$$\text{for all } u \in \Omega(r). \qquad (1.3)$$

holds. Here and below in this chapter

$$\zeta(h) = [\int_{-\eta}^0 \|h(s)\|_{C^n}^2 \, ds]^{1/2}$$

for a vector-valued function h defined on $[-\eta, 0]$.

Let, for example, the condition

$$\|F(t, u_t)\|_{C^n} \leq \sum_{j=0}^m q_j \|u(t - h_j(t))\|_{C^n}^{p_j} \text{ for all } u \in \Omega(r) \qquad (1.4)$$

hold, where $p_j \geq 1$ are numbers, $h_0(t) \equiv 0$, and $h_j(t)$ $(j = 1, ..., m)$ are differentiable functions with the properties

$$1 - \dot{h}_j(t) \geq b_j > 0, \text{ and } 0 \leq h_j(t) < \eta. \qquad (1.5)$$

Here b_j and q_j are constant numbers;

$$t > 0, j = 1, ..., m; \dot{h} \equiv dh/dt.$$

Then as proved below, the following result is true:

Proposition 8.1.2 *Let (1.4) and (1.5) hold. Then for a given positive r, the condition (1.3) is valid with*

$$q = \sum_{j=0}^{m} q_j r^{p_j - 1} b_j^{-1/2}, \ b_0 = 1. \tag{1.6}$$

Notice that the system

$$dx/dt = A_0 x(t) + A_1 x(t - h_1) + ... + A_m x(t - h_m) + F(t, x_t) \tag{1.7}$$

with constant matrices A_k and $0 = h_0 < h_1 < ... < h_m$ takes the shape (1.1) with

$$R(\tau) = A_k \text{ for } h_{k-1} < \tau \le h_k.$$

Let A be an $n \times n$-matrix. Recall that the quantity $g(A)$ and the numbers $\gamma_{n,k}, k = 0, ..., n - 1$ are defined in Section 1.1. Due to Corollary 1.3.7

$$g(A) \le \sqrt{1/2} N(A^* - A). \tag{1.8}$$

Besides, thanks to Corollary 1.3.8

$$g(Ae^{i\theta} + zI) = g(A) \text{ for a real number } \theta \text{ and a complex number } z. \tag{1.9}$$

Denote

$$K(p) = \int_0^{\eta} exp(-ps) dR(s) - Ip \text{ for a complex point } p, \tag{1.10}$$

and

$$G(K(p)) = \sum_{k=0}^{n-1} \frac{g^k(K(p))\gamma_{n,k}}{d^{k+1}(p)},$$

where $d(p)$ is the smallest modulus of eigenvalues of $K(p)$, i.e.

$$d(p) = \min\{|\lambda_k(K(p))| : k = 1, ..., n\}.$$

Put

$$G_0 \equiv \sup\{G(K(\omega i)) : -\infty < \omega < \infty\}.$$

Below we will give a few estimations for G_0. Further, set

$$V(R) \equiv Variation(\|R(\tau)\|_{C^n}) = \int_0^{\eta} \|dR(s)\|_{C^n}, \ 0 \le \tau \le \eta,$$

and

$$H(\Phi) = 2[\|\Phi(0)\|_{C^n} + V(R) \int_{-\eta}^0 \|\Phi(s)\|_{C^n} ds] \sqrt{G_0(V(R)G_0 + 1)}.$$

Finally, denote

$$\Lambda(\Phi) \equiv (1 - qG_0)^{-1} \sqrt{2(V(R) + q)(\zeta(\Phi) + H(\Phi))(qG_0\zeta(\Phi) + H(\Phi))}.$$

8.2   Stability conditions

Everywhere below in the present chapter we assume that all the zeros of $det K(p)$ belong to the open left half-plane.

Theorem 8.2.1 *Let the condition*

$$qG_0 < 1 \tag{2.1}$$

hold. Then the zero solution of (1.1) under (1.3) is stable. Any initial condition (1.2) satisfying the inequality

$$\Lambda(\Phi) \le r, \tag{2.2}$$

belongs to the region of attraction of the zero solution. Specifically, if (2.1) and (2.2) hold, then the bound

$$\|x(t)\|_M \equiv \sup_{t \ge 0} \|x(t)\|_{C^n} \le \Lambda(\Phi) \tag{2.3}$$

is true.

For a real ω denote by $z(\omega, q)$ the extreme right (unique positive and simple) zero of the algebraic equation

$$z^n = qP(z, \omega), \tag{2.4}$$

where

$$P(\lambda, \omega) = \sum_{k=0}^{n-1} g^k(K(\omega i)) \gamma_{n,k} \lambda^{n-k-1}.$$

Corollary 8.2.2 *If $z(\omega, q) < d(i\omega)$ for all $\omega \in [-\infty, \infty]$, then the zero solution of (1.1) is stable under the condition (1.3).*

Notice that (2.1) implies the inequality

$$\Lambda(\Phi) \leq \sqrt{2(V(R) + q)}(1 - qG_0)^{-1}(\zeta(\Phi) + H(\Phi)). \tag{2.5}$$

Thanks to (1.9) and the evident relationship

$$g(A) \leq N(A) \leq \sqrt{n}\|A\|_{C^n}$$

we obtain

$$g(K(i\omega)) \leq v_0 \text{ for all real } \omega,$$

where

$$v_0 \equiv \int_0^\eta N(dR(s)) \leq \sqrt{n} V(R).$$

Let us suppose that

$$K(p) = \sum_{k=0}^{m} A_k \exp(-ph_k) - pI, \tag{2.6}$$

where A_k are constant matrices. Besides, $h_0 = 0$ and $h_k > 0, k = 1, ..., m$ are constants. Then we have due to the definition of $g(A)$ and (1.8)

$$g(K(i\omega)) \leq \sqrt{1/2} \sum_{k=0}^{m} N(A_k \exp(-i\omega h_k) - A_k^* \exp(i\omega h_k))$$

for real ω. Hence,

$$g(K(i\omega)) \leq v_1 \equiv \sqrt{1/2} N(A_0 - A_0^*) + \sqrt{2} \sum_{k=1}^{m} N(A_k) \tag{2.7}$$

for all real ω.

We will also prove the following:

Lemma 8.2.3 *The relationship*

$$\inf_{-\infty \leq \omega \leq \infty} d(i\omega) = \inf_{-2V(R) \leq \omega \leq 2V(R)} d(i\omega) \leq V(R)$$

is valid for any function $K(i\omega)$ defined by (1.10) if $R(\tau)$ has a bounded variation.

Thus Lemma 8.2.3 gives the estimation

$$G_0 \leq \sum_{k=0}^{n-1} \frac{v_0^k \gamma_{n,k}}{d_V^{k+1}}, \tag{2.8}$$

where

$$d_V \equiv \inf_{-2V(R) \leq \omega \leq 2V(R)} d(i\omega).$$

In the case (2.6) and due to (2.7), we can write down

$$G_0 \leq G_1 \equiv \sum_{k=0}^{n-1} \frac{v_1^k \gamma_{n,k}}{d_V^{k+1}}. \tag{2.9}$$

Example 8.2.4 *Let us consider in $\mathbf{C}^n$ the system*

$$dx/dt = A_0 x + A_1 x(t-1) + F(t, x_t),$$

where A_0 and A_1 are constant matrices, and F satisfies the condition

$$\|F(t, x_t)\| \leq q_1 \|x\|_{\mathbf{C}^n}^{p_0} + q_2 \|x(t-1)\|_{\mathbf{C}^n}^{p_1}$$

with p_0, $p_1 \geq 1$.

We assume that (1.2) is satisfied with $\eta = 1$.

Take $r = 1$, then thanks to (1.6), the condition (1.3) is satisfied with $q = q_0 + q_1$. Besides,

$$K(p) = A_0 + A_1 e^{-p} - pI.$$

By (2.7) we can write down the inequality

$$g(K(i\omega)) \leq v_1 = \sqrt{1/2}\, N(A_0 - A_0^*) + \sqrt{2}\, N(A_1)$$

for all real ω.

It is clear that for the considered example

$$V(R) = \|A_0\|_{C^n} + \|A_1\|_{C^n}.$$

Let $\lambda_j(K(i\omega))$ be the eigenvalues of the matrix $K(i\omega)$ with a fixed real ω. Put

$$d_V = \inf_{-2V(R) \leq \omega \leq 2V(R)} \min_{j=1,\dots,n} |\lambda_j(K(i\omega))|.$$

In particular, if A_0 and A_1 are triangular matrices, then

$$d_V = \inf_{-2V(R) \leq \omega \leq 2V(R)} \min_{j=1,\dots,n} |a_{jj}^{(0)} + a_{jj}^{(1)} e^{-i\omega} - i\omega I|,$$

where $a_{jj}^{(k)}$ are the diagonal elements of the matrix A_k ; $k = 0, 1$. Exploiting Theorem 8.2.1 and inequality (2.9), we can assert that the condition

$$qG_1 \equiv q \sum_{k=0}^{n-1} \frac{v_1^k \gamma_{n,k}}{d_V^{k+1}} < 1$$

yields the stability of the zero solution.

To estimate the region of attraction we invoke inequality (2.5). It gives

$$\Lambda(\Phi) \leq \Lambda_1(\Phi) \equiv (1 - qG_1)^{-1}(\zeta(\Phi) + H_1(\Phi))\sqrt{2(V(R) + q)},$$

where

$$H_1(\Phi) \equiv 2[\|\Phi(0)\|_{C^n} + V(R) \int_{-1}^{0} \|\Phi(s)\|_{C^n} ds]\sqrt{G_1(V(R)G_1 + 1)}.$$

Theorem 8.2.1 implies that Φ belongs to the region of attraction of the zero solution if $\Lambda_1(\Phi) \leq 1$. Moreover, the bound $\|x\|_M \leq \Lambda_1(\Phi)$ is true.

8.3 Absolute stability

Let us suppose that in (1.3) $\Omega(r) = C^n$. That is, the inequality

$$\|F(., u_t)\|_L \equiv [\int_0^\infty \|F(t, u_t)\|_{C^n}^2 dt]^{1/2} \leq q(\|u\|_L + \zeta(u)) \qquad (3.1)$$

for all $u \in L^2([-\eta, \infty), C^n)$ holds.

Definition 8.3.1 *Suppose that (3.1) holds. Then the zero solution of equation (1.1) is said to be L^2-absolutely stable in the class of non-linearities (3.1) if there are constants C_0, C_1 independent of the concrete form of F (but dependent on q) such that the inequality*

$$\|x\|_L \leq C_0\zeta(\Phi) + C_1\|\Phi(0)\|_{C^n} \tag{3.2}$$

holds for each solution $x(t)$ of (1.1) with the initial condition (1.2).

Theorem 8.3.2 *Let the condition (2.1) hold, then the zero solution of (1.1) is L^2-absolutely stable in the class of non-linearities (3.1). Specifically, for any solution $x(t)$ of (1.1) with the initial condition (1.2), inequality (3.2) is valid with:*

$$C_0 = \frac{2}{1 - qG_0};$$

$$C_1 = \frac{1 + 2M\sqrt{G_0(MG_0 + 1)}}{1 - qG_0}.$$

Corollary 8.3.3 *Let $z(\omega, q)$ be the extreme right (unique positive and simple) zero of the algebraic equation (2.4). If $z(\omega, q) < d(i\omega))$ for all $\omega \in [-\infty, \infty]$, then the zero solution of (1.1) is L^2-absolutely stable in the class of non-linearities (3.1).*

Example 8.3.4 *Consider in $\mathbf{C}^n$ the following system*

$$dx/dt = Ax + F(t, x_t), \tag{3.3}$$

where A is a constant matrix, and F satisfies the condition (3.1).

According to (1.7) the system (3.3) can be written in the form (1.1). Besides, $K(p) = A - pI$. By (1.9) the equality $g(K(p)) \equiv g(A)$ is valid. Let

$$\alpha(A) = \max\{Re\lambda_k(A) : k = 1, ..., n\} < 0.$$

It is easy to see that

$$d(K(iw)) = d(A - Iwi) \geq |\alpha(A)|$$

for all real w. Therefore,

$$G_0 \leq G_2 \equiv \sum_{k=0}^{n-1} \frac{g^k(A)\gamma_{n,k}}{|\alpha(A)|^{k+1}}.$$

If $qG_2 < 1$, then the system (3.4) under the condition (1.3) is L^2-absolutely stable due to Theorem 8.3.2.

If A is a Hermitian matrix, then $g(A) = 0$ and $G_2 = 1/|\alpha(A)|$.

According to Corollary 8.3.3 this result can be reformulated in the following way: let $z(q)$ be the extreme right-hand root of the algebraic equation

$$z^n = q \sum_{k=0}^{n-1} g^k(A)\gamma_{n,k} z^{n-k-1}.$$

If the matrix $A + z(q)I$ is a Hurwitz one, then the zero solution of the system (3.3) under (3.1) is L^2-absolutely stable.

Example 8.3.5 *Consider the system*

$$\dot{y}_1(t) + a_{11}y_1(t) + a_{12}y_2(t) + a_{13}y_3 + b_1 y(t-1) = f_1(y(t), y(t-1), t),$$
$$\dot{y}_2(t) + a_{22}y_2(t) + a_{23}y_3(t) + b_2 y_2(t-1) = f_2(y(t), y(t-1), t),$$
$$\dot{y}_3(t) + a_{33}y_3(t) + b_3 y_3(t-1) = f_3(y(t), y(t-1), t),$$

where $y = (y_1, y_2, y_3)$, and a_{kk} are positive numbers. Besides, the scalar valued functions f_j defined on $\mathbf{C}^6 \times [0, \infty)$ satisfy the conditions

$$|f_j(u, h, t)| \leq q_0 \sqrt{(u_1^2 + u_2^2 + u_3^2)/3} + q_1 \sqrt{(h_1^2 + h_2^2 + h_3^2)/3}$$

for every $u = (u_k), h = (h_k) \in \mathbf{C}^3, t \geq 0$.

In this case

$$K(p) = -A - B exp(-p) - pI,$$

where

$$A = (a_{jk})_{j,k=1}^n; \quad B = diag[b_k]_{k=1}^n$$

with

$$a_{jk} = 0 \text{ for } j > k; \ (j, k = 1, ..., n). \tag{3.4}$$

Notice that $g(A) = N(V)$, where V is the nilpotent (super-triangular) part of A (see Section 1.3). This means that for a triangular matrix $A = (a_{jk})$ with the entries (3.4)

$$g(A) = \sqrt{\sum_{1 \leq j < k \leq n} |a_{jk}|^2}. \tag{3.5}$$

Due to (1.6) condition (1.3) is satisfied with $q = q_0 + q_1$. By the equality (3.5) we have

$$g(K(p)) = \sqrt{a_{12}^2 + a_{13}^2 + a_{23}^2} = g(A).$$

Besides,

$$d(p) = \min_{k=1,2,3} |p + a_{kk} + b_k exp(-p)|.$$

It is clear that for the considered system

$$V(R) \leq \sqrt{\sum_{1 \leq k \leq j \leq 3} |a_{jk}|^2} + max_{1 \leq j \leq 3} |b_j|.$$

Besides, $z(\omega, q)$ is the extreme right-hand root of the equation

$$z^3 = q(z^2 + g(A)z + \frac{y^2(A)}{2}).$$

I.e. it independs of ω: $z(\omega, q) = z_1(q)$. If

$$z(q) < |i\omega + a_{kk} + b_k exp(-i\omega)| \text{ for } k = 1, 2, 3$$

and all real ω, then due to Corollary 8.3.3 the zero solution of the considered system is L^2-absolutely stable. This means that the inequality

$$min_{k=1,2,3}(a_{kk} - |b_k|) > z(q)$$

ensures the absolute stability.

8.4 Proofs of assertions presented in Sections 8.2 and 8.3

Proof of Proposition 8.1.2: For a given r, the condition (1.4) implies the inequality

$$\|F(t, u_t)\|_{C^n} \leq \sum_{k=0}^{m} q_j r^{p_j - 1} \|u(t - h_j(t))\|_{C^n} \text{ for every } u \in \Omega(r).$$

Hence
$$\|F(t,u_t)\|_L \le \sum_{k=0}^{m} q_j r^{p_j-1}\|u(t-h_j(t))\|_L.$$

The relation (1.5) gives us the inequality
$$b_j\|u(t-h_j(t))\|_L^2 \le \int_0^\infty \|u(t-h_j(t))\|_{C^n}^2(1-\dot h(t))dt.$$

That is,
$$b_j\|u(t-h_j(t))\|_L^2 \le \int_{-h_j(0)}^\infty \|u(s)\|_{C^n}^2 dt =$$
$$\|u\|_L^2 + \int_{-h_j(0)}^0 \|u(s)\|_{C^n}^2 dt.$$

Consequently,
$$\|u(t-h_j(t))\|_L \le \frac{\|u\|_L + \zeta(u)}{\sqrt b}$$

for $j=1,...,m$. We thus get the stated result.□

In order to prove Theorem 8.2.1 we consider the linear equation
$$\dot u = \int_0^\eta dR(s)u(t-s) + f(t) \text{ for } t \ge 0 \tag{4.1}$$

with $f \in L^2([0,\infty),\mathbf{C}^n)$ and the zero initial condition
$$u(t)=0 \text{ for } -\eta \le t \le 0, \text{ and } u(+0)=0. \tag{4.2}$$

First, we will prove a few lemmata.

Lemma 8.4.1 *Any solution $u(t)$ of the equation (4.1) with the zero initial condition (4.2) satisfies the inequality*
$$\|u\|_L \le \hat b_0\|f\|_L,$$

where
$$\hat b_0 = sup\{\|K^{-1}(i\omega)\|_{C^n} : -\infty < \omega < \infty\}.$$

Proof: It is well known (Hale, 1977), (Myshkis, 1972) that a solution of (4.1) with the zero initial condition (4.2) admits the representation
$$u(t) = -\frac{1}{2\pi}\int_{-\infty}^\infty [K(is)]^{-1}exp(ist)\hat f(is)ds, \tag{4.3}$$

where $\hat{f}(p)$ is the Laplace transform of $f(t)$. Thanks to Parseval's equality $\hat{f} \in L^2((-i\infty, i\infty), \mathbf{C}^n)$ and the equality

$$\frac{1}{2\pi} \int_{-\infty}^{\infty} \|[K(is)]^{-1} \hat{f}(is)\|_{\mathbf{C}^n}^2 ds = \|u\|_L^2.$$

holds. Therefore,

$$\|u\|_L^2 \leq \hat{b}_0^2 \frac{1}{2\pi} \int_{-\infty}^{\infty} \|\hat{f}(is)\|_{\mathbf{C}^n}^2 dw = \hat{b}_0^2 \|f\|_L^2.$$

As claimed.$\square$

Corollary 8.4.2 *Let the inequality $G_0 < \infty$ hold, then the bound*

$$\|u\|_L \leq G_0 \|f\|_L$$

is valid for any solution $u(t)$ of the equation (4.1) with the initial condition (4.2).

In fact, let

$$R_\lambda(A) \equiv (A - \lambda I)^{-1}$$

be the resolvent of an $n \times n$-matrix A. Due to Theorem 1.2.2

$$\|R_\lambda(A)\|_{\mathbf{C}^n} \leq \sum_{k=0}^{n-1} \frac{g^k(A)\gamma_{n,k}}{\rho^{k+1}(A, \lambda)} \text{ for all regular } \lambda, \tag{4.4}$$

where $\rho(A, \lambda)$ is the distance between the spectrum of A and a complex point λ. From (4.4) it follows that for all $w \in (-\infty, \infty)$

$$\|[K(iw)]^{-1}\|_{\mathbf{C}^n} \leq G(K(wi)) = \sum_{k=0}^{n-1} \frac{g^k(K(wi))\gamma_{n,k}}{d^{k+1}(iw)}. \tag{4.5}$$

Hence, $\hat{b}_0 \leq G_0$. Now Lemma 8.4.1 implies the stated result.$\square$

Lemma 8.4.3 *Let a solution $u(t)$ of (4.1) with the zero initial condition (4.2) belong to $L^2([0, \infty), \mathbf{C}^n)$. Then the inequality*

$$\|\dot{u}\|_L \leq V(R)\|u\|_L + \|f\|_L$$

is true.

Proof: From (4.1) it follows that

$$\|\dot{u}\|_L \le I_1 + \|f\|_L,$$

where

$$I_1 = [\int_0^\infty \| \int_0^\eta dR(\tau)u(t-\tau)d\tau\|_{C^n}^2 dt]^{1/2}.$$

We have

$$I_1 \le \int_0^\eta \|dR(\tau)\|_{C^n} [\sup_{-\eta \le \tau \le 0} \int_0^\infty \|u(t-\tau)d\tau\|_{C^n}^2 dt]^{1/2}.$$

Thanks to (4.2)

$$\int_0^\infty \|u(t-\tau)d\tau\|_{C^n}^2 dt = \|u\|_L^2.$$

That is,

$$I_1 \le V(R)\|u\|_L.$$

This gives the assertion of the lemma.□

Lemma 8.4.4 *Let $G_0 < \infty$. Then any solution $y(t)$ of the homogeneous equation*

$$dy/dt = \int_0^\eta dR(s)y(t-s) \text{ for } t \ge 0 \qquad (4.6)$$

with the initial condition

$$y(t) = \Phi(t) \text{ for } -\eta \le t < 0, \text{ and } y(+0) = \Phi(0) \qquad (4.7)$$

satisfies the inequality

$$\|y\|_L \le H(\Phi).$$

Proof: It is well known (Hale, 1977), (Myshkis, 1972) that a solution of (4.6) with the initial condition (4.7) admits the representation

$$y(t) = -\frac{1}{2\pi} \int_{-\infty}^\infty [K(i\omega)]^{-1} exp(i\omega t)\{\Phi(0)+$$

$$\int_0^\eta dR(s) \int_{-s}^0 \Phi(z)exp[-i\omega(z+s)]dz\}d\omega. \qquad (4.8)$$

Set for a positive constant b

$$\psi(p) = (p+b)^{-1}\{\Phi(0) + \int_0^\eta dR(s) \int_{-s}^0 \Phi(z)exp[-p(z+s)]dz\}.$$

Hence,

$$y(t) = -\frac{1}{2\pi} \int_{-\infty}^\infty [K(i\omega)]^{-1}exp(i\omega t)(i\omega + b)\psi(i\omega)d\omega.$$

We can write down

$$y(t) = \dot{w} + bw, \tag{4.9}$$

where

$$w(t) = -\frac{1}{2\pi} \int_{-\infty}^\infty [K(i\omega)]^{-1}exp(i\omega t)\psi(i\omega)d\omega.$$

According to (4.3) $w(t)$ is the solution of the equation

$$\dot{w}(t) = \int_0^\eta dR(s)w(t-s) + h(t) \tag{4.10}$$

with the zero initial condition

$$w(t) = 0 \text{ for } t \le 0 \text{ and } w(+0) = 0.$$

Here

$$h(t) = \frac{1}{2\pi} \int_{-\infty}^\infty e^{i\omega t}\psi(i\omega)d\omega$$

is the original for $\psi(p)$. Corollary 8.4.2 and the Parseval equality imply:

$$\|w\|_L^2 \le G_0^2\|h\|_L^2 = G_0^2\frac{1}{2\pi} \int_{-\infty}^\infty \|\psi(i\omega)\|_{C^n}^2 d\omega. \tag{4.11}$$

Omitting simple calculations, we can get

$$\|\psi(i\omega)\|_{C^n}^2 \le (\omega^2 + b^2)^{-1}(\|\Phi(0)\|_{C^n} + V(R)\int_{-\eta}^0 \|\Phi(z)\|_{C^n}dz)^2.$$

Setting

$$m(\Phi) \equiv \|\Phi(0)\|_{C^n} + V(R)\int_{-\eta}^0 \|\Phi(z)\|_{C^n}dz,$$

we obtain

$$\|\psi(i\omega)\|_{C^n}^2 \le (\omega^2 + b^2)^{-1}m^2(\Phi).$$

Taking into account that

$$\int_{-\infty}^{\infty} (\omega^2 + b^2)^{-1} d\omega = 2\pi b^{-1},$$

we arrive at the inequality

$$\|h\|_L^2 = \frac{1}{2\pi} \int_{-\infty}^{\infty} \|\psi(i\omega)\|_{C^n}^2 \, d\omega \le m^2(\Phi) b^{-1}.$$

Relation (4.11) implies the inequality

$$\|w\|_L \le \frac{G_0 m(\Phi)}{\sqrt{b}}.$$

Due to Lemma 8.4.3

$$\|\dot{w}\|_L \le V(R)\|w\|_L + \|h\|_L \le \frac{(V(R)G_0 + 1)m(\Phi)}{\sqrt{b}}.$$

Now (4.9) gives

$$\|y\|_L \le \|\dot{w}\|_L + b\|w\|_L \le \frac{(V(R)G_0 + 1 + bG_0)m(\Phi)}{\sqrt{b}}.$$

Taking

$$b = G_0^{-1}(V(R)G_0 + 1),$$

we conclude that

$$\|y\|_L \le 2\sqrt{G_0(V(R)G_0 + 1)}\, m(\Phi) = H(\Phi).$$

As claimed. $\Box$

Lemma 8.4.5 *Let a function h and its derivative $\dot{h}$ belong to the space $L^2([0, \infty), C^n)$. Then*

$$\|h\|_M \equiv \sup_{t \ge 0} \|h(t)\|_{C^n} \le \sqrt{2\|h(t)\|_L \|\dot{h}(t)\|_L}.$$

Proof: Indeed, using the identity

$$\|h(t)\|_{C^n}^2 = -\int_t^{\infty} \frac{d\|h(s)\|_{C^n}^2}{ds} \, ds,$$

we can assert that

$$\|h(t)\|^2_{C^n} = -2 \int_t^\infty \|h(s)\|_{C^n} \frac{d\|h(s)\|_{C^n}}{dt}.$$

Due to Schwarz's inequality

$$\|h(t)\|^2_{C^n} \leq 2\|h\|_L \|\dot{h}\|_L,$$

since

$$\left|\frac{d\|h(t)\|_{C^n}}{dt}\right| \leq \|\dot{h}\|_{C^n}.$$

Thus, we arrive at the result.□

Lemma 8.4.6 *Assume that (1.2), (1.3) and (2.1) hold. Besides, suppose that a solution $x(t)$ of (1.1) belongs to $\Omega(r)$. Then the estimation*

$$\|x\|_L \leq (1 - qG_0)^{-1}(qG_0\zeta(\Phi) + H(\Phi)) \qquad (4.12)$$

is true.

Proof: We can rewrite (1.1) in the form (4.1) with $f(t) = F(t, x_t)$. That is,

$$x(t) = u(t) + y(t),$$

where $u(t)$ is the solution of (4.1) with the initial zero condition (4.2) and $y(t)$ is the solution of (4.6) with the initial condition (4.7). Due to (1.3)

$$\|f\|_L \leq q(\|x\|_L + \zeta(\Phi)).$$

Corollary 8.4.2 implies the relation

$$\|x\|_L \leq G_0 q(\|x\|_L + \zeta(\Phi)) + \|y\|_L.$$

Now (2.1) ensures the estimation

$$\|x\|_L \leq (1 - qG_0)^{-1}(qG_0\zeta(\Phi) + \|y\|_L).$$

It remains to use Lemma 8.4.4. □

Lemma 8.4.7 *Assume that the conditions (1.2), (1.3), and (2.1) hold. Besides, suppose that the solution $x(t)$ of (1.1) belongs to $\Omega(r)$. Then the estimation*

$$\|\dot{x}\|_L \le (V(R) + q)(\|x\|_L + \zeta(\Phi))$$

is true.

Proof: Due to (1.1) and (1.3)

$$\|\dot{x}\|_L \le \sqrt{\int_0^\infty (\int_0^\eta \|dR(s)\|_{C^n} \|x(t-s)\|_{C^n})^2 dt} + q(\|x\|_L + \zeta(\Phi)).$$

It is simple to check that

$$\int_0^\infty (\int_0^\eta \|dR(s)\|_{C^n} \|x(t-s)\|_{C^n})^2 dt \le V^2(R) \sup_{0 \le s \le \eta} \int_0^\infty \|x(t-s)\|_{C^n}^2 dt.$$

Take into account that

$$\int_0^\infty \|x(t-s)\|_{C^n}^2 dt \le \|x\|_L^2 + \int_{-\eta}^0 \|\Phi(z)\|_{C^n}^2 dz$$

$$\text{for } \eta \ge s \ge 0.$$

Then

$$\sqrt{\int_0^\infty (\int_0^\eta \|dR(s)\|_{C^n} \|x(t-s)\|_{C^n})^2 dt} \le V(R)(\|x\|_L + \zeta(\Phi)).$$

This proves the result. $\square$

Lemma 8.4.8 *Assume that under (1.3), the inequality (2.1) holds. Besides, suppose that a solution $x(t)$ of (1.1) with the initial condition (1.2) belongs to $\Omega(r)$. Then the estimation (2.3) is true.*

Proof: Lemmas 8.4.5 and 8.4.7 imply

$$\|x\|_M \le \sqrt{2\|\dot{x}\|_L \|x\|_L} \le \sqrt{2(V(R) + q)(\|x\|_L + \zeta(\Phi))\|x\|_L}.$$

Now Lemma 8.4.6 ensures the inequality

$$\|x\|_M \le$$

$$(1 - qG_0)^{-1} \sqrt{2(V(R) + q)(\zeta(\Phi) + H(\Phi))(qG_0\zeta(\Phi) + H(\Phi))} = \Lambda(\Phi).$$

As claimed.□

Proof of Theorem 8.2.1: In the case $r = \infty$ the theorem is valid due to Lemma 8.4.8. Consider the equation

$$dx/dt = \int_0^\eta dR(\tau)x(t - \tau) + F_1(t, x_t) \text{ for } t \geq 0, \qquad (4.13)$$

where F_1 satisfies

$$[\int_0^\infty \|F_1(t, u_t)\|_{C^n}^2 dt]^{1/2} \leq q(\|u\|_L + \zeta(u))$$

$$\text{for all } u \in L^2([-\eta, \infty); C^n) \qquad (4.14)$$

and

$$F(t, u_t) = F_1(t, u_t) \text{ for } u \in \Omega(r). \qquad (4.15)$$

The function F_1 satisfying (4.14) and (4.15) can be constructed due to the well-known Theorem 1.1 from the book (Krasnoselskii and Zabreiko, 1984, p.2) and due to the condition (1.3). Thus, for a solution $x(t)$ of (4.13) Lemma 8.4.8 gives the estimation (2.3). The condition (2.2) guarantees that the solution of (4.13) lies in $\Omega(r)$. But according to (4.15) it is the solution of (1.1), simultaneously. This proves the result.□

Proof of Corollary 8.2.2: Dividing (2.3) by z^n and taking into account that $z(w, q)$ is a root of (2.3), we have

$$\sum_{k=0}^{n-1} \frac{g^k(K(wi))\gamma_{n,k}}{z^{k+1}(w, q)} = 1. \qquad (4.16)$$

If $z(w, q) < d(iw)$ for all real w, then the condition (2.1) is satisfied. Thus, the result follows from (4.16) and Theorem 8.2.1.□

Proof of Lemma 8.2.3: For any eigenvalue $\lambda_j(K(i\omega))$ of $K(i\omega)$ with a fixed ω it can be written

$$\lambda_j(K(i\omega)) = (K(i\omega)h_j(i\omega), h_j(i\omega)),$$

where $h_j(\omega) \in C^n$ is the corresponding eigenvector with

$$\|h_j(\omega)\|_{C^n} = 1$$

and $(.,.)$ is the scalar product in $\mathbf{C}^n$. Similarly,

$$\overline{\lambda}_j(K(i\omega)) = (K^*(i\omega)h_j(i\omega), h_j(i\omega)).$$

Thus

$$\Re\lambda_j(K(i\omega)) = (K_R(i\omega)h_j(i\omega), h_j(i\omega)),$$

$$\Im\lambda_j(K(i\omega) = (K_I(i\omega)h_j(i\omega), h_j(i\omega)), \tag{4.17}$$

where

$$K_R(i\omega) = (K^*(i\omega) + K(i\omega))/2$$

and

$$K_I(i\omega) = (K(i\omega) - K^*(i\omega))/2i.$$

But

$$K_R(i\omega) = \frac{1}{2}(\int_0^\eta e^{-is\omega}dR^*(s) + \int_0^\eta e^{is\omega}(i\omega s)dR(s)).$$

That is,

$$|Re\lambda_j(K(i\omega))| \le \|K_R\|_{C^n} \le V(R). \tag{4.18}$$

Besides,

$$K_I(i\omega) = 1/2i(\int_0^\eta e^{-is\omega}dR^*(s) - \int_0^\eta e^{is\omega}(i\omega s)dR(s)) - \omega I$$

Taking into account (4.17), we can easily check that $\Im\lambda_j(i\omega)$ have zeros. Thus,

$$\min_{-\infty \le \omega \le \infty} |\lambda_k(K(i\omega)| \le |\Re\lambda_k(K(i\omega_0)|,$$

where ω_0 is a zero of $\Im\lambda_k(K(i\omega))$. Now (4.18) implies

$$\min_{-\infty \le \omega \le \infty} |\lambda_k(K(i\omega))| \le V(R). \tag{4.19}$$

On the other hand,

$$\|K(i\omega)h_j(i\omega)\|_{C^n} = \|(\int_0^\eta exp(i\omega s)dR(s) - Ii\omega)h_j(i\omega)\|_{C^n}.$$

That is,

$$\|K(i\omega)h_j(i\omega)\|_{C^n} \ge |\omega| - \|\int_0^\eta exp(i\omega s)dR(s)h_j(i\omega)\|_{C^n}$$

$$\geq |\omega| - V(R)$$

if $|\omega| \geq V(R)$. Consequently,

$$|\lambda_j(K(i\omega))| = \|K(i\omega)h_j(i\omega)\|_{C^n} \geq V(R)$$

if $|\omega| \geq 2V(R)$. This and (4.19) mean that the minimum of $|\lambda_j(K(i\omega)|$ is attained on the segment $|\omega| \leq 2V(R)$. As claimed. $\square$

8.5 On asymptotical behavior of solutions of a class of retarded systems

8.5.1 Introduction and statement of the result

Let us consider in a Euclidean space $\mathbf{C}^n$ with the Euclidean norm $\|.\|$ the equation

$$dx/dt = Ax + F(t, x_t) \text{ for all } t \geq 0, \tag{5.1}$$

where A is a constant $n \times n$ matrix, F maps $[0, \infty) \times C((-\eta, \infty), \mathbf{C}^n)$ with some positive $\eta < \infty$ into $\mathbf{C}^n$ with the following property: there is a positive number $r \leq \infty$ such that

$$\|F(t, u_t)\| \leq \sum_{j=0}^{m} q_j \|u(t - h_j)\| \text{ for all } u \in \Omega(r), \tag{5.2}$$

where h_j are non-negative numbers such that $\eta = max_{k=1,\ldots,m} h_k$, and

$$\Omega(r) = \{u \in C([-\eta, \infty), \mathbf{C}^n) : \|u(t)\| \leq r, \ t \in (-\eta, \infty)\},$$

again.

In this section we derive stability conditions for (5.1) under (5.2). These conditions make Theorem 8.2.1 in the case considered more precise and more simple for applications. Besides, we will obtain for equation (5.2) an estimate for Lyapunov's exponents.

Again take the initial condition in the form

$$x(t) = \Phi(t) \text{ for } -\eta \leq t < 0, \text{ and } x(+0) = \Phi(0), \tag{5.3}$$

where $\Phi(t)$ is a given continuous vector-valued function.

We will exploit Corollary 1.2.3 which gives the estimate :

$$\|exp(At)\| \leq e^{\alpha(A)t} \sum_{k=0}^{n-1} \frac{g^k(A)}{\sqrt{k!}} t^k, \ t \geq 0, \tag{5.4}$$

assuming that $\alpha(A) = \max_k Re\alpha_k(A) < 0$. Denote

$$\mu = \max_{t \geq 0} e^{\alpha(A)t} \sum_{k=0}^{n-1} \frac{g^k(A)}{\sqrt{k!}} t^k \tag{5.5}$$

and

$$\theta = \sum_{k=0}^{n-1} \frac{g^k(A)}{|\alpha(A)|^{k+1}}.$$

Besides, put

$$q = \sum_{j=0}^{m} q_j,$$

and

$$\zeta_1(\Phi) = \sum_{j=0}^{m} q_j \int_{-h_j}^{0} \|\Phi(z)\| dz.$$

Clearly,

$$\zeta_1(\Phi) \leq q \int_{-\eta}^{0} \|\Phi(z)\| dz.$$

Theorem 8.5.1 *Let $q\theta < 1$, then the zero solution of (5.1) under (5.2) is stable. Moreover, any initial function Φ satisfying*

$$(1 - q\theta)^{-1} \mu \ [\|\Phi(0)\| + \zeta_1(\Phi)] < r \tag{5.6}$$

belongs to the region of attraction of the zero solution, and for a solution $x(t)$ with the initial condition (5.3), the estimate

$$\|x(t)\| \leq (1 - q\theta)^{-1} \mu \ [\|\Phi(0)\| + \zeta_1(\Phi)], \ t \geq 0 \tag{5.7}$$

is true.

The proofs are given in the next subsection.

Corollary 8.5.2 *Let $z(q, \eta)$ be the extreme right (positive) root of the scalar equation*

$$1 = qe^{-(\alpha(A)+z)\eta} \sum_{k=0}^{n-1} \frac{g^k(A)}{\sqrt{k!}z^{k+1}}, \tag{5.8}$$

and let the conditions (5.2), (5.6) and $z(q, \eta) < |\alpha(A)|$ be fulfilled. Then for a solution $x(t)$ of (5.1) with the initial condition (5.3), the estimate

$$\|x(t)\| \le c_0 exp[(\alpha(A) + z(q, \eta))t], \ t \ge 0 \ (c_0 = const)$$

is true.

8.5.2   Proofs

Proof of Theorem 8.5.1: The equation (5.1) is equivalent to the following one

$$x(t) = e^{At}x(0) + \int_0^t e^{A(t-s)}F(t, x_t(s))ds.$$

By (5.4)

$$\|x(t)\| \le e^{\alpha(A)t}p(t, A)\|x(0)\| + \int_0^t e^{\alpha(A)(t-s)}p(t-s, A)\|F(t, x_t(s))\|ds, \tag{5.9}$$

where

$$p(t, A) \equiv \sum_{k=0}^{n-1} \frac{g^k(A)t^k}{\sqrt{k!}}.$$

Due to (5.9) and (5.2) solutions of (5.1) continuously depend on the initial values. Thanks to (5.6) $\|x(0)\| < r$ and therefore there is a sufficiently small t_0 such that $\|x(t)\| < r$, $t \le t_0$. According to (5.2) for $t \le t_0$

$$J(x, t) \equiv \int_0^t e^{\alpha(A)(t-s)}p(t-s, A)\|F(t, x_t(s))\|ds \le$$

$$\int_0^t e^{\alpha(A)(t-s)}p(t-s, A) \sum_{k=0}^m q_j\|x(s-h_j)\|ds$$

$$= \sum_{k=0}^m q_j \int_{-h_j}^{t-h_j} e^{\alpha(A)(t-z-h_j)}p(t-z-h_j, A)\|x(z)\|dz.$$

Due to (5.3) and (5.5)

$$\sum_{k=0}^{m} q_j \int_{-h_j}^{0} e^{\alpha(A)(t-z-h_j)} p(t-z-h_j, A) \|x(z)\| dz \le$$

$$\mu \sum_{k=0}^{m} q_j \int_{-h_j}^{0} \|\Phi(z)\| dz = \mu \zeta_1(\Phi).$$

Thus,

$$J(x,t) \le \sum_{k=0}^{m} q_j \int_{0}^{t-h_j} e^{\alpha(A)(t-z-h_j)} p(t-z-h_j, A) \|x(z)\| dz + \mu \zeta_1(\Phi) \le$$

$$\sup_{s \le t_0} \|x(s)\| \sum_{k=0}^{m} q_j \int_{0}^{t-h_j} e^{\alpha(A)(t-z-h_j)} p(t-z-h_j, A) dz + \mu_1 \zeta(\Phi) \le$$

$$\sup_{s \le t_0} \|x(s)\| \sum_{k=0}^{m} q_j \int_{0}^{\infty} e^{\alpha(A)s} p(s) ds + \mu \zeta_1(\Phi), \ t \le t_0. \qquad (5.10)$$

The integration gives

$$\int_{0}^{\infty} e^{\alpha(A)s} p(s) ds = \sum_{k=0}^{n-1} \frac{g^k(A)}{\sqrt{k!} |\alpha(A)|^{k+1}} = \theta.$$

From this and from (5.10) it follows that

$$J(x,t) \le q\theta \sup_{s \le t_0} \|x(s)\| + \mu \zeta_1(\Phi).$$

Now (5.9) implies

$$\sup_{t \le t_0} \|x(t)\| \le \mu \|\Phi(0)\| + q\theta \sup_{s \le t_0} \|x(s)\| + \mu \zeta_1(\Phi).$$

Since $q\theta < 1$, we have

$$\|x(t)\| \le (1 - \theta q)^{-1} \mu [\|\Phi(0)\| + \zeta_1(\Phi)], \ t \le t_0.$$

Thanks to (5.6) we can extend this result to $[0, \infty)$. The proof is complete. $\square$

For a positive $\nu < |\alpha(A)|$ put

$$q_\nu \equiv \sum_{j=0}^{m} q_j e^{\nu h_j}, \text{ and } \theta_\nu \equiv \sum_{k=0}^{n-1} \frac{g^k(A)}{\sqrt{k!} |\alpha(A) + \nu|^{k+1}}.$$

Lemma 8.5.3 *Let (5.6) hold, and for some positive $\nu \le |\alpha(A)|$, the condition*

$$q_\nu \theta_\nu < 1 \tag{5.11}$$

be fulfilled. Then for a solution $x(t)$ of (5.1) with the initial condition (5.3), the estimate

$$\|x(t)\| \le c_0 exp[(-\nu)t], \ c_0 = const, \ t \ge 0,$$

is valid.

Proof: Make in (5.1) the substitution

$$x(t) = y(t)e^{-\nu t}. \tag{5.12}$$

We get

$$\dot{y} = (A + I\nu)y + F_1(t, y_t) \text{ for all } t \ge 0, \tag{5.13}$$

where

$$F_1(t, y_t) = e^{\nu t} F(t, e^{-\nu t} \hat{y}_t)$$

with the notation

$$\hat{y}_t = y(t + \tau)e^{-\nu \tau} \ (-\eta \le \tau \le 0).$$

Employing (5.2), we obtain

$$\|F(t, e^{-\nu t} y_t)\| \le e^{-\nu t} \sum_{k=0}^{m} q_j e^{\nu h_j} \|y(t - h_j)\| \text{ if } e^{-\nu t} y(t) \in \Omega(r).$$

That is

$$\|F_1(t, y_t)\| \le \sum_{k=0}^{m} q_j e^{h_j \nu} \|y(t - h_j)\| \text{ if } e^{-\nu t} y \in \Omega(r).$$

Due to Corollary 1.3.8, $g(A) = g(A + \alpha I)$ for any number α. If (5.11) holds, then due to Theorem 8.5.1 a solution $y(t)$ of (5.13) is bounded. It remains to take into account (5.12). $\square$

   Proof of Corollary 8.5.2 : Evidently, $q_\nu \le q e^{\nu \eta}$. Thus, the condition (5.11) is ensured by the inequality

$$q e^{\nu \eta} \sum_{k=0}^{n-1} \frac{g^k(A)}{|\alpha(A) + \nu|^{k+1} \sqrt{k!}} < 1.$$

Or

$$qe^{-(\alpha(A)+z)\eta} \sum_{k=0}^{n-1} \frac{g^k(A)}{\sqrt{k!}z^{k+1}} < 1, \tag{5.14}$$

where $z = -\alpha(A) - \nu > 0$

The function in the right side of (5.8) decreases as $z > 0$ increases, therefore, the condition (5.14) is satisfied if $\nu = -\alpha(A) - z(q,\eta) - \epsilon > 0$ for a sufficiently small ϵ. Now the result is due to Lemma 8.3.5, because ϵ is arbitrarily small.

8.6   Notes

The great variety of types of processes and disturbances has led to numerous mathematical notions of stability, among them stability for hereditary processes (Abrahamson and Infante, 1975), (Agathokis and Foda, 1989), (Barmish and Shi, 1989), (Boese, 1991), (Chiasson, 1988), (Datko, 1986), (Diblasio, Kunish, and Sinestrari, 1985) etc.

The stability of nonlinear systems with delay has been discussed by many authors (see e.g. (Kolmanovskii and Myshkis, 1992), (Hale, 1977) and references given therein). The basic method of the analysis of stability is the Lyapunov functionals method. By this method many very strong results are obtained. But finding Lyapunov's functionals is usually difficult.

In the present chapter a new approach is proposed. It allows us to avoid the construction of Lyapunov's functionals in appropriate situations.

The absolute stability of differential-delay systems for other classes of nonlinearities has been examined in many papers (see e.g. (Rasvan, 1975), (Likhtarnikov and Yakubovich, 1983), etc). Usually, the frequency criteria for ordinary equations are extended.

The contents of Sections 8.2-8.5 are adapted from Gil' (1994) (see also the book by Kolmanovskii and Myshkis (1992, p.111)).

Chapter 9

Absolute Stability of Solutions of Systems of Volterra Integral Equations

This chapter is devoted to stability of systems of Volterra integral equations.

In Sections 9.2 an absolute (global) stability criterion is formulated.

In Section 9.3 the proofs are presented.

Relevant examples are given in Section 9.4.

9.1 Notation

Consider in a Euclidean space $\mathbf{C}^n$ the following nonlinear Volterra integral equation

$$x(t) - \int_0^t K(t-s)x(s)ds + F(x,t) = f(t), \ t \geq 0, \qquad (1.1)$$

where $K(t)$ is a matrix-valued function satisfying the conditions listed below, $f \in L^2([0,\infty), \mathbf{C}^n)$ and $F(x,t)$ maps $L^2([0,\infty), \mathbf{C}^n) \times [0,\infty)$ into

$L^2([0,\infty),\mathbf{C}^n)$ under the condition

$$\|F(x,t)\|_L \le q\|x\|_L \text{ for all } x \in L^2([0,\infty),\mathbf{C}^n) \qquad (1.2)$$

with a positive constant q. As usual $L^2([0,\infty),\mathbf{C}^n)$ is the space of functions defined on $[0,\infty)$ with values in $\mathbf{C}^n$ and the norm

$$\|x\|_L \equiv [\int_0^\infty \|x(t)\|_{\mathbf{C}^n}^2\,dt]^{1/2}.$$

Lemma 9.1.1 *Let the mapping $F(x,t) : L^2([0,\infty),\mathbf{C}^n) \times [0,\infty) \to \mathbf{C}^n$ be represented by the formula*

$$F(x,t) = \int_0^t Q(t,s,x(s))ds,$$

where the function $Q(t,s,h) : [0,\infty) \times [0,\infty) \times \mathbf{C}^n \to \mathbf{C}^n$ satisfies the inequality $\|Q(t,s,h)\|_{\mathbf{C}^n} \le q_1(t,s)\|h\|_{\mathbf{C}^n}$ $(t,s \ge 0, h \in \mathbf{C}^n)$, and the scalar-valued function $q_1(t,s)$ has the property

$$q_2 = \sqrt{\sup_{\tau \ge 0} \int_0^\tau q_1(\tau,t) \int_t^\infty q_1(s,t)dt\,ds} < \infty$$

Then (1.2) holds with $q = q_2$.

In particular, if $q_1(t,s) = q_1(t-s)$, then by Lemma 9.1.1 we easily obtain condition (1.2) with

$$q = \int_0^\infty q_1(s)ds.$$

Further, let, for instance,

$$F(x,t) = \int_0^t Q(t,s,x(s))ds + \psi_2(x(t),t),$$

where $Q(t,s,x)$ is the same as above, and the function ψ_2 maps $\mathbf{C}^n \times [0,\infty)$ into $\mathbf{C}^n$ with the property

$$\|\psi_2(h,t)\|_{\mathbf{C}^n} \le q_3\|h\|_{\mathbf{C}^n} \text{ for all } h \in \mathbf{C}^n,$$

with a constant q_3. Then due to Lemma 9.1.1 condition (1.2) holds with $q = q_3 + q_2$. Here q_2 is the same as above.

A solution of (1.1) is a continuous vector-valued function x(t) satisfying this equation for all t ≥ 0.

We assume the existence of solutions, although a priori estimates obtained below make it possible to prove the existence by the traditional scheme with the help of the Leray-Schauder principle.

Definition 9.1.2 *Suppose that (1.2) holds. Then equation (1.1) is said to be L^2 -absolute stable in the class of non-linearities (1.2) if there is a constant $\eta(q)$ independent of the concrete form of F (but dependent on q) such that*

$$\|x\|_L \leq \eta(q)\|f\|_L$$

for each solution x of (1.1).

We will exploit Theorem 1.2.2 which implies the following estimate for the resolvent of an n × n-matrix A

$$\|R_\lambda(A)\|_{C^n} \leq \sum_{k=0}^{n-1} \frac{g^k(A)\gamma_{n,k}}{\rho^{k+1}(A,\lambda)} \tag{1.3}$$

for any regular λ. Recall that $\rho(A,\lambda)$ is the distance between the spectrum of A and a complex point λ. The numbers $\gamma_{n,k}$, $k=0,...,n-1$ and $g(A)$ are introduced in Section 1.1.

Due to Corollaries 1.3.7 and 1.3.8 for any operator A in C^n and any complex point z the relations

$$g^2(A) \leq N^2(A-A^*)/2 \tag{1.4}$$

and

$$g(A+zI_{C^n}) = g(A) \tag{1.5}$$

are true.

9.2   Statement of the result

Let $H(p)$ be the Laplace transform of $K(t)$, and p the dual variable. Denote by $d(p)$ the least modulus of eigenvalues of the matrix $I - H(p)$:

$$d(p) = \min_{k=1,...,n} |1 - \lambda_k(H(p))|.$$

Set

$$w = \inf_{-\infty < s < \infty} \sum_{k=0}^{n-1} \frac{g^k(H(is))\gamma_{n,k}}{d^{k+1}(is)}. \tag{2.1}$$

Theorem 9.2.1 *Let under (1.2) the inequality*

$$wq < 1 \tag{2.2}$$

hold. Then (1.1) is L^2-absolutely stable, and the bound

$$\|x\|_L \leq w (1 - wq)^{-1}\|f\|_L \tag{2.3}$$

is true for any solution x of (1.1).

The proofs are presented in the next section.

Denote by $z(q,p)$ the extreme right root of the algebraic equation

$$z^n = q \sum_{k=0}^{n-1} g^k(H(p))\gamma_{n,k} z^{n-k-1} \quad (p \in \mathbf{C}). \tag{2.4}$$

It is not hard to check that $z(q,p)$ is a unique positive and simple root of this equation.

Corollary 9.2.2 *Let under (1.2), the inequality*

$$d(is) > z(q, is) \tag{2.5}$$

hold for all $s \in [-\infty, \infty]$. Then the zero solution of (1.1) is L^2-absolutely stable.

9.3 Proofs

We need the following simple result.

Lemma 9.3.1 *Let a positive scalar-valued function $K(t,s) \in L^2([a,b] \times [a,b])$ satisfy the conditions $K(t,s) = K(s,t)$ $(s,t \in [a,b])$ and*

$$k \equiv sup_{t \in [a,b]} \int_a^b K(t,s)ds < \infty.$$

Then a linear operator A defined by

$$Au(t) = \int_a^b K(t,s)u(s)ds$$

is bounded in $L^2(a,b)$ and its norm in this space subordinates the bound $\|A\|_L \le k$.

Proof: Since A is a positive operator in $L^2(a,b)$, we can apply the well-known estimate $d(A) \le k$ for the spectral radius $d(A)$ cf. (Krasnoselskii, 1964). But A is a selfadjoint operator. Therefore, $\|A\|_L = d(A)$. This proves the stated result.$\square$

Lemma 9.3.2 *Let a positive scalar-valued function $W(t,s) \in L^2([a,b] \times [a,b])$ satisfy the condition*

$$k_1 \equiv \sqrt{\sup_{\tau \in [a,b]} \int_a^b \int_a^b W(\tau,t)W(s,t)dt\,ds} < \infty.$$

Then a linear operator A defined by

$$Au(t) = \int_a^b W(t,s)u(s)ds$$

is bounded in $L^2(a,b)$, and its norm in this space subordinates the bound $\|A\|_L \le k_1$.

Proof: Put $M = AA^*$. We can write down

$$Mu(\tau) = \int_a^b W(\tau,t) \int_a^b W(s,t)u(s)ds\,dt.$$

Lemma 9.3.1 clearly forces

$$\|M\|_L = \sup_{\tau \in [a,b]} \int_a^b \int_a^b W(\tau,s)W(s,t)ds\,dt = k_1^2.$$

But

$$\|A\|_L \le \sqrt{\|M\|_L}.$$

We thus get the result.$\square$

Corollary 9.3.3 *Let a linear operator A be defined in $L^2(0,\infty)$ by*

$$Au(t) = \int_0^t W_1(t-s)u(s)ds,$$

where the scalar-valued function $W_1(t)$ has the property

$$k_2 = \int_0^\infty |W_1(s)|ds < \infty.$$

Then the norm of A in $L^2(0,\infty)$ satisfies the inequality $\|A\|_L \le k_2$.

Indeed, due to Lemma 9.3.2

$$\|A\|_L^2 \le \sup_{\tau \ge 0} \int_0^\tau \int_t^\infty |W_1(\tau-t)|\,|W_1(s-t)|dtds.$$

Since

$$\int_t^\infty |W_1(s-t)|ds \le k_2, \ t \ge 0$$

and

$$\int_0^\tau |W_1(\tau-t)|dt \le k_2, \ \tau \ge 0,$$

we arrive at the stated result.

Proof of Lemma 9.1.1: Under the hypothesis of Lemma 9.1.1

$$\|F(x,t)\|_L^2 \le \int_0^\infty [\int_0^t q_1(t,s)\|x(s)\|_{C^n}ds]^2 dt \le$$

$$\|A_1\|_L \int_0^\infty \|x(s)\|_{C^n}^2 ds,$$

where A_1 is a Volterra integral operator with the kernel $q_1(t,s)$, $t \ge s \ge 0$ defined in the space of scalar-valued functions $L^2(0,\infty)$. Due to Lemma 9.3.2

$$\|A_1\|_L^2 \le \sup_{\tau \ge 0} \int_0^\infty \int_0^\infty q_1(\tau,t)q_1(s,t)dtds.$$

Since $q_1(t,s) = 0$ for $t \le s$, we arrive at the result. $\square$

To prove Theorem 9.2.1 consider the linear equation

$$y(t) - \int_0^t K(t-s)y(s)ds = \phi(t), \tag{3.1}$$

where ϕ belongs to the space $L^2([0,\infty), C^n)$.

Lemma 9.3.4 *Let $w < \infty$. Then any solution $y(t)$ of (3.1) satisfies the inequality*

$$\|y\|_L \leq w\|\phi\|_L.$$

Proof: We apply the Laplace transformation to (3.1). It implies the equation

$$\overline{y}(p) - H(p)\overline{y(p)} = \overline{\phi}(p),$$

where $\overline{y}(p)$ and $\overline{\phi}(p)$ are the Laplace transforms of $y(t)$ and of $\phi(t)$, respectively. Hence,

$$\overline{y(p)} = (I - H(p))^{-1}\overline{\phi}(p)$$

and therefore,

$$\|\overline{y}(p)\|_{C^n} \leq \|(I - H(p))^{-1}\|_{C^n}\|\overline{\phi}(p)\|_{C^n}. \tag{3.2}$$

According to (1.5)

$$g(I - H(p)) = g(H(p)).$$

Due to (1.3)

$$\|(I - H(p))^{-1}\|_{C^n} \leq \sum_{k=0}^{n-1} \frac{g^k(H(p))\gamma_{n,k}}{d^{k+1}(p)}.$$

By (3.2) we obtain the inequality

$$\|\overline{y}(is)\|_{C^n} \leq w\|\overline{\phi}(is)\|_{C^n}.$$

Consequently,

$$\int_{-\infty}^{\infty} \|\overline{y}(is)\|_{C^n}^2\, ds \leq w^2 \int_{-\infty}^{\infty} \|\overline{\phi}(is)\|_{C^n}^2\, ds.$$

Now, the Parseval equality ensures the result.□

    Proof of Theorem 9.2.1: Set in (3.1)

$$\phi(t) = F(x,t) + f(t).$$

Then (3.1) coincides with (1.1). Taking into account (1.2), we have

$$\|\phi\|_L \leq q\|x\|_L + \|f\|_L.$$

Lemma 9.3.4 gives:

$$\|x\|_L \le w\|\phi\|_L \le w(q\|x\|_L + \|f\|_L).$$

Due to (2.2) this implies the bound (2.3) which ensures the absolute stability. □

Proof of Corollary 9.2.2: Dividing (2.4) by z^n and taking into account that $z(q,p)$ is its root, we get for $p = is$ with a real s

$$1 = q \sum_{k=0}^{n-1} \frac{g^k(H(is))\gamma_{n,k}}{z^{k+1}(q,is)}.$$

Condition (2.5) implies inequality (2.2). Now Theorem 9.2.1 yields the stated result. □

9.4   Examples

Example 9.4.1   *Consider the equation (1.1) with $n = 2$:*

$$K(t) = (K_{jk}(t)); \ j,k = 1,2.$$

Let $H(p) = (H_{jk}(p))$ be the Laplace transform of $K(t)$, then

$$\lambda_{1,2}(H(p)) = \frac{1}{2} Tr H(p) \pm \sqrt{\frac{(TrH(p))^2}{4} - \det H(p)}.$$

Further,

$$d(p) = \min_{k=1,2} |1 - \lambda_k(H(p))|.$$

According to (1.4)

$$g(H(is)) \le \sqrt{1/2} N(H^*(is) - H(is)).$$

Thus, if

$$q \left[\frac{1}{d(is)} + \frac{\sqrt{1/2} N(H^*(is) - H(is))}{d^2(is)} \right] < 1$$

for all $s \in [-\infty, \infty]$, then (1.1) under (1.2) is absolutely L^2-stable due to Theorem 9.2.1.

Example 9.4.2 *Let $K(t) = v(t)B$ where B is a constant $n \times n$-matrix and $v(t)$ is a scalar-valued function from $L^2[0,\infty)$.*

In the considered example

$$H(p) = \overline{v}(p)B,$$

where $\overline{v}(p)$ is the Laplace transform of $v(t)$. According to the definition of $g(A)$ we can write down

$$g(H(is)) = |\overline{v}(is)|g(B) \text{ for every real } s.$$

Besides,

$$d(is) = \min_{k=1,...,n} |1 - \overline{v}(is)\lambda_k(B)|.$$

Now we can apply Theorem 9.2.1.

Example 9.4.3 . *Consider (1.1) assuming that $K(t)$ is a triangular matrix for all $t > 0$.*

That is, we assume $K_{jk}(t) = 0$ if $j > k$ $(j, k = 1, 2, ..., n)$. It is clear that the Laplace transform $H(p)$ is a triangular matrix as well. Therefore $1 - H_{jj}(p)$ are eigenvalues of $I - H(p)$, and

$$d(p) = \min_{j=1,...,n} |1 - H_{jj}(p)|.$$

Take into account that $g(A) = N(V)$ for any matrix A, where V is the nilpotent part of A (see Section 1.3). Then

$$g(H(p)) = [\sum_{1 \le j < k \le n} |H_{jk}(p)|^2]^{1/2}.$$

Now we can apply Theorem 9.2.1.

9.5 Notes

The absolute (global) stability of nonlinear Volterra integral equations has been discussed by many authors (see for instance (Gripenberg, Londen, and Staffans, 1990) and references given therein). Basic methods for the analysis of the global stability are the frequency domain methods.

The results obtained in this chapter supplement the frequency domain methods in the case of systems with many nonlinearities under the above assumptions.

The contents of this chapter are new.

Chapter 10

Stability of Semilinear Parabolic Systems

In this chapter a class of semilinear evolution equations is analyzed. It contains various parabolic and integrodifferential systems. Estimates for solution norms are derived. As consequence of these estimations, stability conditions are obtained.

In Section 10.1 the well-known results about sectorial operators and analytical semigroups are collected.

Section 10.2 deals with semilinear evolution equations in a Banach space.

The main result of the chapter is Theorem 10.3.1. It is presented in Section 10.3 and gives stability conditions for equations on a tensor product of Hilbert spaces.

In Section 10.4 particular cases of Theorem 10.3.1 are considered.

Stability of a second order parabolic system is investigated in Section 10.5 and stability of a fourth order parabolic system is considered in Section 10.6.

10.1 Sectorial operators and analytical semigroups

Let X be a Banach space with a norm $\|.\|$.

A linear operator A in X *is sectorial* if A is a closed densely defined linear operator such that there exist constants $\phi \in (0, \pi/2), M_0 \geq 1$ and a real b such that the sector

$$S_{b,\phi} \equiv \{\lambda \in \mathbf{C} : \phi \leq |arg(\lambda - b)| \leq \pi, \ \lambda \neq b\}$$

belongs to the resolvent set of A, and

$$\|(I\lambda - A)^{-1}\| \leq M_0/|\lambda - b| \text{ for all } \lambda \in S_{b,\phi}.$$

Recall that I is the unit operator.

An analytic semigroup on a Banach space X is a family of continuous linear operators on X, $\{T(t)\}, t \geq 0$ satisfying

(i) $T(0) = I$, $T(t)T(s) = T(t + s)$ for $t, s \geq 0$,

(ii) $T(t)x \rightarrow x$ as $t \rightarrow +0$ for each $x \in X$ in the sense of the strong topology,

(iii) $t \rightarrow T(t)x$ is real analytic on $0 < t < \infty$.

The infinitesimial generator of this semigroup is defined by

$$Ax = \lim_{t \to 0+} \frac{1}{t}(T(t)x - x),$$

its domain $D(A)$ consisting of all $x \in X$ for which this limit exists in the sense of the strong topology.

We usually write $T(t) = e^{At}$.

It is well-known (see, for example, Henry (1981, pp. 20-23)) that if A is a sectorial operator, then $-A$ generates an analytic semigroup e^{-At}.

If A is a sectorial operator on X with $\Re\sigma(A) > 0$, then for any $\nu > 0$, one can define

$$A^{-\nu} = \frac{1}{\Gamma(\nu)} \int_0^\infty t^{\nu-1} e^{-At} dt,$$

where $\Gamma(\nu)$ is the gamma function evaluated at ν. The operator $A^{-\nu}$ is a bounded linear operator on X which is one-to-one and satisfies

$$A^{-\nu}A^{-\mu} = A^{-(\nu+\mu)} \text{ for } \mu, \nu > 0$$

(Henry, 1981, p. 25).

One can now define A^ν for $\nu > 0$ as the inverse of $A^{-\nu}$. The operator A^ν, $\nu > 0$ is closed and densely defined. Define $A^0 = I$. Also, if $\Re\sigma(A) > \delta > 0$, then for any $\nu \geq 0$ there is a constant $C_\nu < \infty$ such that

$$\|A^\nu e^{At}\| \leq C_\nu t^{-\nu} e^{-\delta t} \; ; t > 0.$$

If A is a sectorial operator in X, then there is a number a such that $A_1 = A + aI$ has $Re\sigma(A_1) > 0$. If we define $X^\nu = D(A_1^\nu)$ with the graph norm

$$\|x\|_{X^\nu} = \|A_1^\nu x\|, \; x \in X^\nu,$$

then X^ν is a Banach space with norm $\|.\|_{X^\nu}$. Furthermore if X^ν is defined by different a with the above property, then the norms are equivalent and so the dependence on a is suppressed. Besides, $X^0 = X$, and X^μ is a dense subspace of X^ν for any $\mu \geq \nu \geq 0$.

10.2 Stability of semilinear evolution equations in a Banach space

10.2.1 Definitions

In a Banach space X with a norm $\|.\|$, let us consider the equation

$$du/dt = Au + F(u,t), \; t \geq 0 \qquad (2.1)$$

where A is a closed linear operator with the domain $D(A)$ dense in X, and F maps $D(A) \times [0,\infty)$ into X, and $F(0,t) \equiv 0$. Besides A generates a strongly continuous semigroup e^{At} which is bounded on $[0,\infty)$:

$$\chi \equiv \sup_{t \geq 0} \|e^{At}\| < \infty.$$

Definition 10.2.1 *A solution of (2.1) over $[0, \infty)$ is a continuous function $u : [0, \infty) \to X$, $u(0) = u_0$ satisfying the equation*

$$u(t) = e^{tA} u(0) + \int_0^t e^{(t-s)A} F(u(s), s) ds \qquad (2.2)$$

for all $t \geq 0$.

That is, we considered the mild solutions (Henry, 1981, p. 53).

Definition 10.2.2 *Let Z be a subspace of X with a norm $\|.\|_Z$. We will say that the zero solution of (2.1) is (Z)-stable if for any sufficiently small $\epsilon > 0$ there is a constant $\delta > 0$ such that the condition $\|u_0\|_Z \leq \delta$ for any given $u_0 \in Z$ implies the existence of a solution $u(t) : [0, \infty) \to Z$ with $u(0) = u_0$, and the relationship*

$$\|u(t)\|_Z \leq \epsilon \text{ for any } t \geq 0$$

is valid.

Definition 10.2.3 *We will say that the zero solution of equation (2.1) is (Z)-asymptotically stable if it is (Z)-stable and there is a constant δ such that the condition $\|u(0)\|_Z \leq \delta$ implies the relationship $\|u(t)\|_Z \to 0$ as $t \to \infty$. Besides, the set*

$$B(\delta) \equiv \{h \in Z : \|h\|_Z \leq \delta\}$$

will be called a region of attraction of the zero solution.

If, besides, $\delta = \infty$, that is, if $B(\delta) = Z$, then we will say that the zero solution of (2.1) is globally asymptotically (Z)-stable.

10.2.2 Statement of the result

Let $\sigma(A)$ be the spectrum of an operator A, and $\alpha(A) \equiv \sup \Re \sigma(A)$.

Our assumptions in this section are as follows:

(i) There is a linear sectorial operator $-G$ acting in X and commuting with A such that $D(G) = D(A)$ and $\alpha(G) < 0$.

(ii) For some $r > 0$ and $\nu \in [0, 1)$ the condition

$$\|F(h_1, t) - F(h_2, t)\| \leq q(t, r)\|G^\nu(h_1 - h_2)\|$$

for all $h_1, h_2 \in \Omega(r, G^\nu)$ and $t \geq 0$ $\qquad$ (2.3)

is satisfied, where $q(t, r)$ is a non-negative bounded measurable scalar-valued function of t and

$$\Omega(r, G^\nu) = \Omega(X, r, G^\nu) \equiv \{h \in D(G^\nu) : \|G^\nu h\| \leq r\}.$$

(iii) For any $\mu \in [0, 1)$, the relation

$$\eta(\mu, t) \equiv \|G^\mu e^{At}\| \in L^1(0, \infty) \qquad (2.4)$$

is valid.

Define on the set $D(G^\nu)$ the graph norm

$$\|h\|_\nu = \|G^\nu h\| \quad (h \in D(G^\nu)).$$

Denote the Banach space with that norm by Y^ν.

Theorem 10.2.4 *Let conditions (i)-(iii) and*

$$b_\nu(r) \equiv \sup_{t \geq 0} \int_0^t \eta(\nu, t - s)q(s, r)ds < 1 \qquad (2.5)$$

hold, then the zero solution of (2.1) is (Y^ν)-asymptotically stable. Any initial vector $u(0) = u_0 \in Y^\nu$ satisfying the inequality

$$\chi\|u_0\|_\nu \leq r(1 - b_\nu(r)) \qquad (2.6)$$

belongs to the region of attraction of the zero solution. Moreover, under (2.6) the estimate

$$\|u(t)\|_\nu \leq \chi\|u(0)\|_\nu(1 - b_\nu(r))^{-1} \leq r \text{ for all } t \geq 0 \qquad (2.7)$$

is true.

The proof of this theorem is presented in the next subsection.

Let $q(t, r)$ satisfy the inequality $q(t, r) \leq q_0(t)\theta(r)$, where $q_0(t)$ is a function bounded on $[0, \infty)$ and independent of r, and $\theta(r)$ is a non-negative non-decreasing continuous scalar-valued function of a non-negative argument r independent of t. That is,

$$\|F(h_1, t) - F(h_2, t)\| \leq q_0(t)\theta(r)\|h_1 - h_2\|_\nu$$

$$\text{for all } h_1, h_2 \in \Omega(r, G^\nu), t \geq 0. \tag{2.8}$$

Put

$$\xi_\nu \equiv \sup_{t \geq 0} \int_0^t \eta(\nu, t - s)q_0(s)\,ds.$$

We have $b_\nu(r) \leq \theta(r)\xi_\nu$.

Corollary 10.2.5 *Let conditions (2.8), (2.4) and*

$$\theta(0)\xi_\nu < 1 \tag{2.9}$$

hold, then the zero solution of (2.1) is (Y^ν)-asymptotically stable. If, in addition, the scalar equation

$$x = \chi\|u_0\|_\nu + x\theta(x)\xi_\nu \tag{2.10}$$

has a positive root, then the initial vector $u(0) = u_0 \in Y^\nu$ belongs to the region of attraction of the zero solution. Moreover, the estimate

$$\|u(t)\|_\nu \leq x(\theta, u_0) \text{ for all } t \geq 0 \tag{2.11}$$

is true, where $x(\theta, u_0)$ is the smallest positive root of (2.10).

10.2.3   Proof of Theorem 10.2.4 and its corollary

Lemma 10.2.6 *Let the conditions*

$$\|F(h, t)\| \leq q(t, r)\|h\|_\nu \text{ for all } h \in \Omega(r, G^\nu) \tag{2.12}$$

and (2.5) hold. Let there also be a solution $u : [0, \infty) \to Y^\nu$ of (2.1) with an initial vector u_0 satisfying (2.6). Then this solution satisfies estimate (2.7).

Proof: Put $T(t) = e^{At}$. Since the solution exists, according to (2.4) we can multiply (2.2) by G^ν. Take into account that G commutes with A, consequently it does with $T(t)$. We easily obtain

$$\|G^\nu u(t)\| \leq \|T(t)G^\nu u(0)\| + \int_0^t \|G^\nu T(t-s)F(u(s),s)\|ds.$$

It is clear that $\chi \geq 1$. So that condition (2.6) implies that $\|G^\nu u_0\| \leq r$. Since $T(0) = I$ and $T(t)$ is a continuous function, we can assert that

$$\|G^\nu u(t)\| \leq r \text{ for } t \leq t_0$$

if t_0 is sufficiently small. Now, using (2.12), we arrive at the inequality

$$\|G^\nu u(t)\| \leq \|T(t)G^\nu u(0)\| + \int_0^t \|G^\nu T(t-s)\|q(s,r)\|G^\nu u(s)\|ds$$

for $t \leq t_0$. This implies

$$\|u(t)\|_\nu \leq \chi\|u(0)\|_\nu + \int_0^t \eta(\nu,t-s)q(s,r)\|u(s)\|_\nu\,ds \text{ for } t \leq t_0.$$

Hence,

$$\sup_{t \leq t_0} \|u(t)\|_\nu \leq \chi\|u(0)\|_\nu + \sup_{s \leq t_0} \|u(s)\|_\nu \sup_{t \geq 0} \int_0^t \eta(\nu,t-s)q(s,r)ds.$$

Thus,

$$\sup_{t \leq t_0} \|u(t)\|_\nu \leq \chi\|u(0)\|_\nu + sup_{s \leq t_0}\|u(s)\|b_\nu(r).$$

Thanks to (2.5)

$$\sup_{t \leq t_0} \|u(t)\|_\nu \leq \chi\|u(0)\|_\nu (1-b_\nu(r))^{-1} \text{ for sufficiently small } t_0. \qquad (2.13)$$

According to (2.6) we can extend this inequality to $[0,\infty)$. As claimed.$\square$

Lemma 10.2.7 *Let (2.3)-(2.5) hold. Let an initial vector $u_0 \in D(A)$ satisfy condition (2.6). Then there exists a unique solution $u : [0,\infty) \to Y^\nu$ of (2.2) with $u(0) = u_0$.*

Proof: For $u : [0, \infty) \to X$ introduce the norm

$$\|u\|_M \equiv \sup_{t \geq 0} \|u(t)\|$$

and put

$$\Omega_M(r, \nu) \equiv \{u : [0, \infty) \to \Omega(r, G^\nu)\}.$$

Define on $\Omega_M(r, \nu)$ the mapping Ψ by

$$\Psi(u)(t) \equiv T(t)u(0) + \int_0^t T(t - s)F(u(s), s)ds.$$

We have due to (2.3)

$$\|\Psi(u_1)(t) - \Psi(u_2)(t)\|_\nu \leq$$

$$\sup_{t \geq 0} \int_0^t \|G^\nu T(t - s)\|q(s, r)\|u_1(s) - u_2(s)\|_\nu \, ds \text{ for } u_1, u_2 \in \Omega_M(r, \nu).$$

Thus

$$\|G^\nu(\Psi(u_1) - \Psi(u_2))\|_M \leq \|G^\nu(u_1 - u_2)\|_M \sup_{t \geq 0} \int_0^t \|G^\nu T(t - s)\|q(s, r)ds.$$

Hence

$$\|G^\nu(\Psi(u_1) - \Psi(u_2))\|_M \leq \|G^\nu(u_1 - u_2)\|_M \, b_\nu(r) \text{ for } u_1, u_2 \in \Omega(r, G^\nu).$$

Taking into account Lemma 10.2.6 and (2.6), we can assert that Ψ maps $\Omega_M(r, \nu)$ into itself. According to (2.5) Ψ is a contraction. Therefore equation (2.2) has a unique solution $u \in \Omega_M(r, \nu)$. As claimed.$\square$

   Proof of Theorem 10.2.4: Substitute in (2.1) $u(t) = u_1(t)exp(-\epsilon t)$ with a sufficiently small positive ϵ .We obtain the equation

$$du_1/dt = (A + \epsilon)u_1 + F_1(u_1, t), \ t \geq 0 \tag{2.14}$$

with the notation

$$F_1(u_1, t) = e^{\epsilon t} F(u_1 e^{-\epsilon t}, t).$$

According to (2.3) the condition

$$\|F_1(h_1, t) - F_1(h_2, t)\| \leq q(t, r)\|G^\nu(h_1 - h_2)\|$$

for all $h_1, h_2 \in \Omega(r, G^\nu)$ and $t \geq 0$

is satisfied. Now Lemma 10.2.7 ensures the solution existence for (2.14) with sufficiently small ϵ. Due to Lemma 10.2.6 $G^\nu u_1(t)$ is uniformly bounded on $[0, \infty)$. Therefore, $G^\nu u(t) = e^{-t\epsilon} G^\nu u_1(t)$ tends to zero as t tends to infinity. The proof is complete.$\square$

Proof of Corollary 10.2.5: Since θ is continuous, there is an $r \leq \infty$ such that due to (2.9) $b_\nu(r) \leq \theta(r)\xi_\nu < 1$. Thanks to Theorem 10.2.4 under (2.6) there is a solution $u \in \Omega_M(r, \nu)$. Now (2.13) implies the inequality

$$\|G^\nu u\|_M \leq \chi \|u_0\|_\nu + \phi(r)\xi_\nu \|G^\nu u\|_M \leq \chi \|u_0\|_\nu + r\phi(r)\xi_\nu, \qquad (2.15)$$

since ϕ does not decrease. Further, taking into account that $x(\phi, u_0)$ is the root of (2.10), we obtain

$$\chi \|u_0\|_\nu + r\phi(r)\xi_\nu \leq x(\theta, u_0)$$

for any $r \leq x(\theta, u_0)$. This and (2.15) prove estimate (2.11).$\square$

10.3 Stability of semilinear evolution equations on a tensor product of Hilbert spaces

10.3.1 Notation

Let H be a tensor product of a Euclidean space $\mathbf{C}^n$ and a separable Hilbert space L, i.e. $H = \mathbf{C}^n \otimes L$ (see Section 6.1). The scalar product in H is defined as

$$(y \otimes h, y_1 \otimes h_1)_H = (y, y_1)_L (h, h_1)_{\mathbf{C}^n} ; y, y_1 \in L, \text{ and } h, h_1 \in \mathbf{C}^n,$$

where $(.,.)_L$ and $(.,.)_{\mathbf{C}^n}$ are the scalar products in L and $\mathbf{C}^n$, respectively. The norms in $L, \mathbf{C}^n$ and H are denoted by $\|.\|_L, \|.\|_{\mathbf{C}^n}$ and $\|.\|_H$, respectively. $I, I_L, I_{\mathbf{C}^n}$ are the identity operators in H, L and $\mathbf{C}^n$, respectively.

Throughout this section S is a normal operator in L with a compact inverse one. That is,

$$S = \sum_{k=1}^{\infty} \lambda_k(S) P_k,$$

where P_k are the orthogonal eigenprojectors of S, and

$$|\lambda_k(S)| \to \infty \text{ as } k \to \infty.$$

Besides, we assume that $\alpha(S) = \max_k \Re \lambda_k(S) < 0$. It is well-known that a normal operator with the spectrum in a half-plane is sectorial (Henry, 1981). So that S is sectorial. Put

$$S_0 = S \otimes I_{C^n}.$$

It is clear that S_0 is normal in H and $\alpha(S) = \alpha(S_0) < 0$.

Define on $D(S_0^\nu)$ the graphic norm $\|h\|_\nu = \|S_0^\nu h\|_H$. Denote the obtained space by H^ν. Besides, $H = H^0$. If $X = H$ and $G = S_0$, then $Y^\nu = H^\nu$. In the sequel $\|.\|_\nu$ is the norm in H^ν (not in Y^ν).

Let now W_k, $k = 1, 2, \ldots$ be a sequence of $n \times n$-matrices with the property

$$0 < \lim_{k \to \infty} |\lambda_k(S)|^{-1} \|W_k\|_{C^n} < \infty. \tag{3.1}$$

Denote

$$D(W(S)) = \{ u \in H : \sum_{k=1}^{\infty} \|(W_k \otimes P_k) u\|_{C^n}^2 < \infty \}$$

and define on $D(W(S))$ an operator $W(S)$ by the formula

$$W(S) = \sum_{k=1}^{\infty} W_k \otimes P_k. \tag{3.2}$$

This means that

$$W(S)u = \sum_j \sum_{k=1}^{\infty} (W_k h_j) \otimes (P_k y_j)$$

for

$$u = \sum_j h_j \otimes y_j \in D(W(S)) \text{ with } y_j \in L \, ; h_j \in C^n.$$

The series converges in the norm of H.

Consider in H equation (2.1) with $A = W(S)$. That is, we consider the equation

$$du/dt = W(S)u + F(u,t), \ t \geq 0, \tag{3.3}$$

where F maps $\Omega(r, S_0^\nu) \times [0, \infty)$ into H with the following properties: $F(0,t) = 0$, and for a positive $r \leq \infty$ and $\nu \in [0,1)$, the inequality

$$\|F(h_1,t) - F(h_2,t)\|_H \leq q(t,r)\|h_1 - h_2\|_\nu$$

$$\text{for all } h_1, h_2 \in \Omega(r, S_0^\nu) \ , t \geq 0 \tag{3.4}$$

holds. Here $q(t,r)$ is the same as above and

$$\Omega(r, S_0^\nu) = \Omega(H, r, S_0^\nu) = \{h \in D(S_0^\nu) : \|h\|_\nu \leq r\}.$$

Below we will show that $W(S)$ generates a strongly continuous semigroup. So that a solution is defined as in the previous section.   Let Q be an n-dimensional operator, $n < \infty$. Recall that quantity $g(Q)$ and the numbers $\gamma_{n,k}$ for $k = 0, ..., n-1$ are defined in Section 1.1.

Due to Corollaries 1.3.7 and 1.3.8, the following relations

$$g^2(Q) \leq N^2(Q - Q^*)/2 \tag{3.5}$$

and

$$g(zI_{C^n} + Q) = g(Q) \text{ for any complex } z \tag{3.6}$$

are true.

10.3.2   Statement of the main result

We will suppose that

$$g_0 \equiv \sup_{j=1,2,...} g(W_j) = \sup_{j=1,2,...} [N^2(W_j) - \sum_{k=1}^{n} |\lambda_k(W_j)|^2]^{1/2} < \infty$$

and

$$\alpha_0 \equiv \sup_{j=1,2,...} \alpha(W_j) < 0.$$

Put

$$\eta(W,\nu,t) \equiv \sup_{j=1,2,\ldots} |\lambda_j(S)|^\nu exp[t\alpha(W_j)] \sum_{k=0}^{n-1} t^k g_0^k \frac{\gamma_{n,k}}{k!}$$

and

$$\chi(W) \equiv \sup_{t\geq 0} \eta(W,0,t).$$

Since $\alpha_0 < 0$, then $\chi(W) < \infty$.

Theorem 10.3.1 *Let the conditions (3.4) and*

$$b_\nu(r,W) \equiv \sup_{t\geq 0} \int_0^t \eta(W,\nu,t-s)q(s,r)ds < 1$$

be fulfilled for a positive $r \leq \infty$ and a $\nu \in [0,1)$. Then the zero solution of equation (3.3) is (H^ν)-asymptotically stable. Besides, any initial vector $u(0) = u_0 \in D(S_0)$ satisfying the inequality

$$\chi(W)\|u_0\|_\nu \leq (1 - b_\nu(r,W))\,r, \tag{3.7}$$

belongs to the region of attraction of the zero solution. Moreover, the estimate

$$\|u(t)\|_\nu \leq \frac{\chi(W)\|u_0\|_\nu}{1 - b_\nu(r,W)} \quad \text{for all } t \geq 0 \tag{3.8}$$

is true .

The proofs are presented in the next section.

Below we will prove that Theorem 10.3.1 is precise.

Corollary 10.3.2 *Let the condition*

$$\|F(h_1,t) - F(h_2,t)\|_H \leq q_0(t)\|h_1 - h_2\|_\nu$$

for all $h_1, h_2 \in H^\nu$,$t \geq 0$

hold with a $\nu \in [0,1)$, and $q_0(t)$ is bounded on $[0,\infty)$. Let besides,

$$b_\nu(\infty,W) \equiv \sup_{t\geq 0} \int_0^t \eta(W,\nu,t-s)q_0(s)ds < 1.$$

Then the zero solution of equation (3.3) is (H^ν)-globally asymptotically stable. Moreover, (3.8) is true with $r = \infty$ for any solution u of (3.3).

Let us suppose that for a fixed $r \leq \infty$ the quantity $q(t,r) = \theta(r)$, is independent of t. Then for $\nu = 0$ the relations

$$b_0(r, W) = \theta(r) \sup_{t \geq 0} \int_0^t \eta(W, 0, t-s) ds = \theta(r) \int_0^\infty \eta(W, 0, s) ds$$

$$= \theta(r) \int_0^\infty exp[s\alpha_0] \sum_{k=0}^{n-1} s^k g_0^k \frac{\gamma_{n,k}}{k!} ds$$

hold. Integration gives

$$b_0(r, W) = \theta(r) \sum_{k=0}^{n-1} \frac{g_0^k \gamma_{n,k}}{|\alpha_0|^{k+1}}. \tag{3.9}$$

Further, let $q(t,r)$ satisfy the inequality $q(t,r) \leq q_0(t)\theta(r)$, where $q_0(t)$ is bounded and independent of r, and $\theta(r)$ is continuous non-decreasing and independent on t. Inequality (3.4) implies

$$\|F(h_1, t) - F(h_2, t)\|_H \leq q_0(t)\theta(r)\|h_1 - h_2\|_\nu \tag{3.10}$$

$$\text{for all } h_1, h_2 \subset \Omega(r, S_0^\nu), t \geq 0.$$

Put

$$\xi_\nu(W) \equiv \sup_{t \geq 0} \int_0^t \eta(\nu, W, t-s) q_0(s) ds,$$

then $b_\nu(r, W) \leq \theta(r)\xi_\nu(W)$.

Corollary 10.3.3 *Let the conditions (3.10) and*

$$\theta(0)\xi_\nu(W) < 1$$

hold, then the zero solution of (3.1) is (H^ν)-asymptotically stable. If, in addition, the scalar equation

$$x = \chi\|u_0\|_\nu + \xi_\nu(W)x\theta(x) \tag{3.11}$$

has a positive root, then the initial vector $u(0) = u_0 \in D(S_0^\nu)$ belongs to the region of attraction of the zero solution. Moreover, the estimate

$$\|u(t)\|_\nu \leq x(\theta, u_0, W) \text{ for all } t \geq 0 \tag{3.12}$$

is true where $x(\theta, u_0, W)$ is the smallest positive root of (3.11).

10.3.3 Proof of Theorem 10.3.1

Lemma 10.3.4 *The operator $W(S)$ under (3.1) generates a strongly continuous semigroup $exp[W(S)t]$ which subordinates the inequality*

$$\|S_0^\nu e^{W(S)t}\| \le \eta(W, \nu, t) \text{ for } t > 0. \tag{3.13}$$

Proof: Since

$$D(S_0) = \{u \in H : \sum_j |\lambda_j(S)|^2 \|(P_j \otimes I_{C^n})u\|_H^2 < \infty\},$$

due to (3.1) we easily obtain the equality $D(W(S)) = D(S_0)$.

Set

$$T(t) = \sum_{j=1}^{\infty} exp[W_j t] \otimes P_j. \tag{3.14}$$

For $u = h \otimes y$ with $y \in L$; $h \in \mathbf{C}^n$ we have

$$S_0^\nu T(t)u = \sum_{j=1}^{\infty} \lambda_j(S)^\nu exp[W_j t]h \otimes P_j y \,, t > 0$$

Since P_j are reciprocally orthogonal,

$$\|T(t)u\|_\nu^2 = \sum_{j=1}^{\infty} |\lambda_j(S)|^{2\nu} \|exp[W_j t]h\|_{C^n}^2 \|P_j y\|_L^2 \,, t > 0.$$

Now, omitting simple calculations, we get

$$\|S_0^\nu W(S)T(t)\|_H = \max_{j=1,2,\ldots} |\lambda_j(S)|^\nu \|exp[W_j t]\|_{C^n} \,, t > 0.$$

Further, we exploit Theorem 1.2.1. It implies

$$\|exp(Qt)\|_{C^n} \le exp[t\alpha(Q)] \sum_{k=0}^{n-1} g^k(Q)t^k \frac{\gamma_{n,k}}{k!}, t \ge 0.$$

for any $n \times n$-matrix Q. We have

$$\|\lambda_j^\nu(S)exp[(W_j)t]\|_{C^n} \le$$

$$|\lambda_j(S)|^\nu exp[t\alpha(W_j)] \sum_{k=0}^{n-1} t^k g^k(W_j) \frac{\gamma_{n,k}}{k!} \le \eta(W, \nu, t).$$

Hence

$$\|T(t)u\|_\nu \le \eta(W,\nu,t)\|u\|_H \text{ for } t > 0.$$

In particular,

$$\|T(t)\|_H \le \eta(W,0,t) \text{ for } t \ge 0.$$

That is, $T(t)$ is bounded. It remains to check that $T(t)$ is a semigroup generated by $W(S)$.

Indeed, according to (3.14) and the equality $D(W(S)) = D(S_0)$ it follows that

$$W(S)T(t)u = \sum_{j=1}^{\infty} (W_j exp[W_j t] \otimes P_j)u$$

for all $u \in D(S_0)$.

On the other hand,

$$\frac{dT(t)u}{dt} = \sum_{j=1}^{\infty} \frac{d}{dt}(exp[W_j t] \otimes P_j)u = \sum_{j=1}^{\infty} W_j exp[W_j t] \otimes P_j u.$$

Thus

$$\frac{dT(t)u}{dt} = W(S)T(t)u.$$

That is, $T(t)$ is a strongly continuous semigroup generated by the operator $W(S)$. $\square$

The assertion of Theorem 10.3.1 follows from Theorem 10.2.4 and Lemma 10.3.4.

Corollary 10.3.3 follows from Theorem 10.3.1, Lemma 10.3.4 and Corollary 10.2.5.

10.4   Particular cases of Theorem 10.3.1

4.1. We begin with a simple case. Let us consider in H the equation

$$\dot{u} = S_0 u + cu = F(u,t), \tag{4.1}$$

where c is a constant $n \times n$-matrix, and $F : D(S_0) \times [0,\infty) \to H$ satisfies the condition

$$\|F(u_1,t) - F(u_2,t)\|_H \le q\|u_1 - u_2\|_H \text{ for all } u_1, u_2 \in D(S_0).$$

That is, (3.4) is satisfied with $\nu = 0$ and $r = \infty$. Since

$$W(S)u = (S \otimes I_{C^n} + I_L \otimes c)u = S_0 u + cu,$$

one can assert that (4.1) has the form (3.3) with

$$W_j = I_{C^n} \lambda_j(S) + c. \tag{4.2}$$

Due to (3.6)

$$g(W_j) = g(\lambda_j(S) I_{C^n} + c) = g(c).$$

Exploiting (3.5), we can write down

$$g(W_j) \le \sqrt{1/2} \, N(c - c^*). \tag{4.3}$$

Furthermore,

$$\alpha(W_j) = \Re \lambda_j(S) + \alpha(c) \le \alpha_0 \text{ for all } j = 1, 2, \dots$$

where $\alpha_0 \equiv \alpha(S) + \alpha(c)$. Besides,

$$\eta(W, 0, t) = e^{\alpha_0 t} \sum_{k=0}^{n-1} g^k(c) \frac{\gamma_{n,k}}{k!} t^k.$$

Put according to (3.9)

$$b_0(\infty, W) = b_0(W) = q \sum_{k=0}^{n-1} g^k(c) \frac{\gamma_{n,k}}{|\alpha_0|^{k+1}}. \tag{4.4}$$

Employing Corollary 10.3.2, we can assert that if $b_0(W) < 1$, then the zero solution of (4.1) is (H)-globally asymptotically stable, and (3.8) holds with $\nu = 0$ and $b_\nu(r, W) = b_0(W)$.

Notice that if c is a normal matrix, then $g(c) = 0$ and $b_0(W) = q/|\alpha_0|$.

4.2. Again consider (4.1), assuming now that S is a selfadjoint negative defined operator and F satisfies the condition

$$\|F(u_1, t) - F(u_2, t)\|_H \le \theta(r) \|S_0^{1/2}(u_1 - u_2)\|_H$$

$$\text{for } u_k \in D(S_0) \text{ with } \|S_0^{1/2} u_k\|_H \le r \,; k = 1, 2. \tag{4.5}$$

That is, (3.4) is satisfied with $\nu = 1/2$ and $q(t,r) = \theta(r)$.

In the considered case W_j is defined by (4.2) and (4.3) is valid. Furthermore,

$$\eta(W, 1/2, t) = \sup_{j=1,2,\ldots} exp[(\lambda_j(S) + \alpha(c))t]|\lambda_j(S)|^{1/2} \sum_{k=0}^{n-1} g^k(c)\frac{\gamma_{n,k}}{k!}t^k .$$

Omitting simple calculations, we obtain

$$\max_{j=1,2,\ldots} e^{\lambda_j(S)t}|\lambda_j(S)|^{1/2} \leq \zeta(t)$$

where

$$\zeta(t) = \begin{cases} exp[-t\alpha(S)]\sqrt{|\alpha(S)|} & \text{if } t \geq 1/(2|\alpha(S)|), \\ \frac{exp[-1/2]}{\sqrt{2t}} & \text{if } t \leq 1/(2|\alpha(S)|). \end{cases}$$

Consequently,

$$\eta(W, 1/2, t) = \zeta(t)exp[\alpha(c)t] \sum_{k=0}^{n-1} g^k(c)\frac{\gamma_{n,k}}{k!}t^k$$

and for $\alpha(S) + \alpha(c) < 0$ we get

$$b_{1/2}(r, W) = \theta(r) \sum_{k=0}^{n-1} g^k(c)\frac{\gamma_{n,k}}{k!} \int_0^\infty \zeta(t)e^{\alpha(c)t}t^k \, dt. \qquad (4.6)$$

This integral is simply calculated. Exploiting Theorem 10.3.1, we can assert that if $b_{1/2}(r, W) < 1$, then the zero solution of (4.1) under (4.5) is $(H^{1/2})$-asymptotically stable, any initial vector u_0 satisfying

$$\chi(W)\|S_0^{1/2}u_0\|_H(1 - b_{1/2}(r, W))^{-1} \leq r,$$

belongs to the region of attraction of the zero solution and (3.8) holds with $\nu = 1/2$.

10.5   Stability of a second order parabolic system

In this and in the next section we will consider simple examples which illustrate the applications of Theorem 10.3.1 and by themselves do not claim to be essential.

Let ω be a bounded domain in a real Euclidean space $\mathbf{R}^l$ with a sufficiently smooth boundary $\partial\omega$. Consider in $\mathbf{C}^n$ the equation

$$\partial u/\partial t = \Delta u + cu + f(u,x,t) \text{ in } \omega \text{ and for all } t \geq 0 \qquad (5.1)$$

with the boundary condition

$$u = 0 \text{ on } \partial\omega \text{ and } t \geq 0.$$

Here $u = u(x,t)$, c is a constant matrix, Δ is the Laplace operator, f maps $\mathbf{C}^n \times \omega \times [0,\infty)$ into $\mathbf{C}^n$ with the property

$$\|f(h_1,x,t) - f(h_2,x,t)\|_{\mathbf{C}^n} \leq q\|h_1 - h_2\|_{\mathbf{C}^n}$$

$$\text{for all } h_1, h_2 \in \mathbf{C}^n, x \in \omega \text{ and } t \geq 0; \; q = const. \qquad (5.2)$$

The space $H = L^2(\omega, \mathbf{C}^n)$ is pertinent for our aim. It is clear that $L^2(\omega, \mathbf{C}^n) = L^2(\omega) \otimes \mathbf{C}^n$. Take

$$D(S) = D(\Delta) \equiv \{y \in L \equiv L^2(\omega) : \Delta y \in L; y = 0 \text{ on } \partial\omega\}.$$

Thus $D(S) = H_0^1(\omega, \mathbf{C}^1) \cap H^2(\omega, \mathbf{C}^1)$. By $H_0^1(\omega, \mathbf{C}^n), H^2(\omega, \mathbf{C}^n)$ the Sobolev spaces are denoted (see Section 6.6.1). Put

$$Sy = \Delta y \text{ for } y \in D(S).$$

That is, S is the Friedrichs extension of the Laplacian operator defined on differentiable finite functions. Taking W_j in the form (4.2), we arrive at the equality

$$\Delta u + cu = W(S)u.$$

Besides, $g(W_j) = g(c)$, the relation (4.3) is true and $\alpha(W(S)) = \alpha(S) + \alpha(c)$ (see Remark 10.5.1 below). It is clear that (5.2) implies (3.4) with $\nu = 0$.

Let $b_0(W)$ be defined by (4.4). Exploiting Corollary 10.3.2, we can assert that if $b_0(W) < 1$, then the zero solution of (5.1) under (5.2) is (H)-globally asymptotically stable. Moreover, (3.8) is true with $\nu = 0$ and $b_\nu(r,W) = b_0(W)$ for any initial vector $u_0 \in D(S_0)$. Observe that in the considered case

$$D(S_0) = H_0^1(\omega, \mathbf{C}^n) \cap H^2(\omega, \mathbf{C}^n).$$

Remark 10.5.1 *In the considered example $S = \Delta$ is selfadjoint. Thus $\alpha(\Delta) = \lambda_1(\Delta)$, where $\lambda_1(\Delta)$ is the extreme right eigenvalue. If ω is a canonical domain (sphere, cylinder, parallelepiped etc), then $\alpha(\Delta)$ is well known. For instance*

$$\alpha(\Delta) = -\pi^2 \sum_{k=1}^{l} \frac{1}{(b_k - a_k)^2}$$

if ω is the parallelepiped $\{a_k \leq x_k \leq b_k, k = 1, ..., l\}$.

If ω is not a canonical we can estimate $\alpha(\Delta)$ by the similar value for a canonical domain which contains ω (Courant, 1962).

10.6 Stability of a fourth order parabolic system

Let ω be a bounded domain in a real Euclidean space $\mathbf{R}^l$ with a sufficiently smooth boundary $\partial\omega$, again. Consider in the space C^n the equation

$$\partial u/\partial t = -\Delta^2 u + cu + f(\Delta u, u_x, u, x, t) \text{ in } \omega \text{ for all } t \geq 0 \qquad (6.1)$$

with a constant matrix c and the boundary condition

$$u = \Delta u = 0 \text{ on } \partial\omega \text{ for all } t \geq 0. \qquad (6.2)$$

The vector-valued function f maps $C^{n(l+2)} \times \omega \times [0, \infty)$ into C^n. By u_x we denote the vector-valued function whose coordinates are the first derivatives of u with respect to all the variables.

Take $H = L^2(\omega, C^n)$.

In order not to overburden the presentation we make the following assumption: for some positive $r \leq \infty$

$$\|f(\Delta h, h_x, h, x, t) - f(\Delta u, u_x, u, x, t)\|_H \leq \theta(r)\|\Delta(h - u)\|_H \qquad (6.3)$$

for all

$$t \geq 0 \text{ and } h, u \in H_0^1(\omega) \cap H^2(\omega); \|\Delta h, \Delta u\|_H \leq r.$$

That is,

$$h, u \in \Omega(r, \Delta) \equiv \{h \in H_0^1(\omega, \mathbf{C}^n) \cap H^2(\omega, \mathbf{C}^n) : \|\Delta h\|_H \leq r\}$$

Below we will give an example of a function satisfying (6.3).

Let S be the Friedrichs extension of the operator S_1 defined by $S_1 u = -\Delta^2 u$ on differentiable functions satisfying (6.2). Evidently, $D(S) \subset H_0^4(\omega)$.

It is simple to show that the operator $S^{1/2}$ satisfies $S^{1/2}u(x) = i\Delta u(x)$ for $u \in H_0^1(\omega, C^1) \cap H^2(\omega, C^1)$. According to (6.3), (4.5) is satisfied. Taking W_j in the form (4.2), we can rewrite (6.1) as (4.1). So that $g(W_j) = g(c)$ (See Section 10.4). Consequently, $\alpha_0 = \alpha(-\Delta^2) + \alpha(c)$ and we can define $b_{1/2}(r, W)$ by (4.6). Besides,

$$\chi(W) = \max_{t \geq 0} exp[t\alpha(-\Delta^2)](1 + g(c)t).$$

About the calculation of $\alpha(-\Delta^2)$ see Remark 10.5.1 above.

Exploiting Theorem 10.3.1, we can assert that if $b_{1/2}(r, W) < 1$, then the zero solution of (6.1) under (6.2) and (6.3) is ($H^{1/2}$)-asymptotically stable, any initial vector u_0 satisfying

$$\chi(W)(1 - b_{1/2}(r, W))^{-1}\|\Delta u_0\|_H \leq r,$$

belongs to the region of attraction of the zero solution, and the inequality

$$\|\Delta u(t)\|_H \leq \chi(W)(1 - b_{1/2}(r, W))^{-1}\|\Delta u_0\|_H \text{ for all } t > 0$$

is true.

Now we consider an example of a function satisfying (6.3). Let $l = 2$, and

$$f(\Delta u, u_x, u, x, t) = \phi(u_{x_1}, u_{x_2}, u),$$

where $\phi(y_1, y_2, y_3)$ is a vector-valued function differentiable with respect to all its variables. Then for a positive $r_0 \leq \infty$ we have

$$\|\phi(y_1, y_2, y_3) - \phi(z_1, z_2, z_3)\|_{C^n} \leq \sum_{k=1}^{3} q_k(r_0)\|y_k - z_k\|_{C^n},$$

$$y_j, z_j \in C^n \; ; \|y_j, z_j\|_{C^n} \le r_0, \; j = 1, 2, 3, \tag{6.4}$$

where

$$q_k(r_0) = \sup\{\|\phi'_{y_k}(y_1, y_2, y_3)\|_{C^n} : \|y_j\|_{C^n} \le r_0 \;, j, k = 1, 2, 3\}.$$

Take into account the well-known representation

$$h(x) = \int_\omega G(x, s) \Delta h(s) d\omega \;, h \in H_0^1(\omega, C^n) \cap H^2(\omega, C^n).$$

Here $G(x, s)$ is the Green function of the Laplace operator defined on $H_0^1(\omega, C^n) \cap H^2(\omega, C^n)$. Using the Schwarz inequality, we arrive at the relationships:

$$\|u(x)\|_{C^n} \le \gamma_3 \|\Delta u\|_H \tag{6.5}$$

and

$$\|u_{x_j}(x)\|_{C^n} \le \gamma_j \|\Delta u\|_H \text{ for each } x \in \omega, \tag{6.6}$$

where

$$\gamma_j = \sup_{x \in \omega} [\int_\omega (\frac{\partial G(x, s)}{\partial x_j})^2 ds]^{1/2} \;, k = 1, 2$$

and

$$\gamma_3 = \sup_{x \in \omega} [\int_\omega G^2(x, s) ds]^{1/2}.$$

These inequalities and the relation $\|\Delta u\|_H \le r$ imply

$$\|u(x)\|_{C^n} \le r_0 \; ; \|u_{x_j}(x)\|_{C^n} \le r_0$$

with $r_0 = r \max_{j=1,2,3} \gamma_j$.

From (6.4), taking into account (6.5) and (6.6), we obtain by the Schwarz inequality that

$$[\int_\omega \|\phi(u_{x_1}(x), u_{x_2}(x), u(x)) - \phi(v_{x_1}(x), v_{x_2}(x), v(x))\|_{C^n}^2 d\omega]^{1/2} \le$$

$$[\sum_{k=1}^3 q_k^2(r_0)]^{1/2} \|\Delta(u - v)\|_H \text{ with } u, v \in \Omega(r, \Delta).$$

Thus, (6.3) is really satisfied with

$$q = \sqrt{\sum_{k=1}^3 q_k^2(r_0)}$$

and $r_0 = r \max_{j=1,2,3} \gamma_j$.

10.7   Notes

7.1. Let S be selfadjoint and all matrices W_j be Hermitian. Let besides, $F(u,t) = qu$. Then $\nu = 0$, according to (3.5)

$$g(W_j) = 0 \text{ and } \eta(W, \nu, t) = e^{\alpha_0 t}.$$

Due to (3.9) $b_0(\infty, W) = q/|\alpha_0|$. Thanks to Theorem 10.3.1 the equation

$$du/dt = W(S)u + qu, \ t \geq 0 \tag{7.1}$$

is stable if $q/|\alpha_0| \ < 1$. On the other hand the operator $A_q = W(S) + qI_H$ is a selfadjoint one with a compact resolvent. Therefore for the stability of (7.1) it is necessary and sufficient that $\alpha(A_q) < 0$. But $\alpha(A_q) = \alpha_0 + q$. Thus, the condition $\alpha(A_q) < 0$ is equivalent to $q/|\alpha_0| < 1$. That is, Theorem 10.3.1 is precise.

7.2. Corollary 10.3.2 supplements the main result from (Krasnosel-skii, Pokrovskii, and Zarubin, 1983). Besides, this corollary permits stability conditions to be obtained for more general nonlinearities F than the criterion suggested in (Yakubovich, 1983) (however, the criterion from (Yakubovich, 1983) under certain restrictions enables us to use properties of F better than Corollary 10.3.2)

7.3. For the application of the well-known results (Henry, 1981) to the problem (6.1), (6.2) one needs to estimate the norm of $A^{1/2}exp(At)$ where A is the operator of the linear part of the problem. The author does not know such an estimate if the matrix c is not Hermitian.

7.4. Evidently, by Theorem 10.3.1 under certain restrictions one can analyze high order parabolic and integral-differential systems, etc.

One of the basic methods for investigation of stability of parabolic equations is the Lyapunov functionals method (see e.g. (Lakshmikantham, Leela, and Martynyuk, 1989), (Lakshmikantham, Matrosov, and Sivasundaram, 1991) and references therein). By this method many very strong results are obtained. But finding the Lyapunov functionals for nonstationary parabolic systems is usually difficult.

Our approach is connected with a method which develops the classical Lyapunov exponent method (Daleckii and Krein, 1971), (Henry, 1981), etc. It is based on estimations for semigroups generated by linear parts of equations or on estimations for corresponding resolvents.

The material of the chapter is based on (Gil', 1984b) and (Gil', 1989b).

Chapter 11

Stability of Volterra Integrodifferential Systems and Applications to Viscoelasticity

This chapter is concerned with the systems of Volterra integrodifferential equations with operator coefficients in a Hilbert space. Some stability conditions are given. These conditions are applied to the stability analysis of a system of two elastic bars attached to one another by a viscoelastic strip.

Section 11.1 contains the relevant notation.

In Section 11.2 the stability conditions are formulated.

The proofs are collected in Section 11.3.

Section 11.4 deals with the mentioned application to a system of two elastic bars.

305

11.1   Notation

Let $H = Y \otimes \mathbf{C}^n$ be a tensor product of a separable Hilbert space Y and an n-dimensional Euclidean space $\mathbf{C}^n$ (see Section 6.1).

Let $\mu_{j_1}^{(1)}$ ($j_1 = 1, \ldots, l_1$) and $\mu_{j_2}^{(2)}$ ($j_2 = 1, \ldots, l_2$) be constant matrices (operators in $\mathbf{C}^n$), and $B_{j_1}^{(1)}$ ($j_1 = 1, \ldots, l_1$) and $B_{j_2}^{(2)}$ ($j_2 = 1, \ldots, l_2$) be normal (in particular, selfadjoint) operators in Y with compact resolvents. These conditions imply that operators $B_{j_1}^{(1)}$ and $B_{j_2}^{(2)}$ have isolated spectra, see (Ahiezer and Glazman, 1981).

Let us consider in H the nonlinear integrodifferential equation

$$\ddot{u}(t) + W_1 u(t) - \int_0^t R(t-s) W_2 u(s) ds = F(t, u(t)), \qquad (1.1)$$

where the superscript dot denotes differentiation with respect to time,

$$W_1 = \sum_{j_1=1}^{l_1} B_{j_1}^{(1)} \otimes \mu_{j_1}^{(1)},$$

$$W_2 = \sum_{j_2=1}^{l_2} B_{j_2}^{(2)} \otimes \mu_{j_2}^{(2)},$$

$R(t)$ is a scalar-valued function (a relaxation kernel) satisfying the conditions

(a) $R(t) \geq 0$, $\dot{R} \leq 0$ for any $t \geq 0$;

(b) $|R| = \int_0^\infty |R(t)| dt < 1$.

Besides, $F(t, u)$ is a nonlinear function mapping $[0, \infty) \times H$ into H and satisfying the conditions listed below.

Introduce a positive selfadjoint operator S in Y, and put $S_0 = S \otimes I_{\mathbf{C}^n}$, and

$$\Omega(r) \equiv \{v \in H : \|v\|_H \leq r\}$$

It is supposed that there are positive constants q and r such that for any continuous function

$$w = w(t) : [0, \infty) \to \Omega(r),$$

the inequality

$$\sup_{0 \le s \le t} \|S_0^{-1} F(s, w(s))\|_H \le q \sup_{0 \le s \le t} \|w(s)\|_H, \ (t \ge 0) \qquad (1.2)$$

holds. We assume also that:

(i) operators $B_{j_1}^{(1)}$ and $B_{j_2}^{(2)}$ commute with one another and with S,

(ii) for any $j_1 = 2, \ldots, l_1$ and $j_2 = 1, \ldots, l_2$

$$D(B_{j_1}^{(1)}) \supseteq D(B_1^{(1)}), \ D(B_{j_2}^{(2)}) \supseteq D(B_1^{(1)}), \ D(S) \supseteq D(B_1^{(1)}).$$

Recall that $D(A)$ denotes the domain of an operator A.

We put $B_0 = B_1^{(1)} \otimes I_{C^n}$. Let H_- be the space dual of $D(B_0)$ via the inner product $(.,.)_H$ of H.

A solution of (1.1) is a function $u(t)$ mapping $[0, \infty)$ into $D(B_0) \cap \Omega(r)$, having for almost all $t \ge 0$ derivatives $\dot{u}(t) \in H$ and $\ddot{u}(t) \in H_-$ which satisfies this equation almost everywhere on $[0, \infty)$.

We assume that for any prescribed initial data

$$u(0) = v_1, \ \dot{u}(0) = v_2 \qquad (1.3)$$

with $v_1 \in D(B_0) \cap \Omega(r)$ and $v_2 \in \Omega(r)$ there exists a solution of (1.1) on $[0, \infty)$.

Introduce the notation

$$A_j^{(1)} = S^{-1} B_j^{(1)}, \ A_j^{(2)} = S^{-1} B_j^{(2)}. \qquad (1.4)$$

Denote by $a_{j,m}^{(1)}$, $a_{j,m}^{(2)}$ and π_m $(m = 1, \ldots)$ the eigenvalues of operators $A_j^{(1)}$, $A_j^{(2)}$ and S, respectively. Introduce the following matrix pencil:

$$\Xi_m(p) = \pi_m^{-1} p^2 I_{C^n} + \sum_{j_1=1}^{l_1} a_{j_1,m}^{(1)} \mu_{j_1}^{(1)} - \hat{R}(p) \sum_{j_2=1}^{l_2} a_{j_2,m}^{(2)} \mu_{j_2}^{(2)}, \qquad (1.5)$$

where $\hat{R}(p)$ is the Laplace transform of $R(t)$ and p is the dual variable.

For a fixed p, denote by

$$\lambda_{k,m}(p) \ (k = 1, \ldots, n)$$

the eigenvalues of matrix $\Xi_m(p)$, and by $\rho_0(\Xi_m(p))$ the minimum of their absolute values

$$\rho_0(\Xi_m(p)) = \min_{k=1,\dots,n} |\lambda_{k,m}(p)|.$$

Recall that the quantity $g(A)$ is defined in Section 1.1.

For a positive constant α we set

$$\Theta(\alpha) = \max_{m=1,\dots} \frac{1}{2\pi} \int_{-\infty}^{\infty} \sum_{k=0}^{n-1} \frac{g^k(\Xi_m(-\alpha+i\omega))d\omega}{\sqrt{k!}\rho_0^{k+1}(\Xi_m(-\alpha+i\omega))} \tag{1.6}$$

and

$$\theta(\alpha) = \frac{\Theta(\alpha)}{\alpha}. \tag{1.7}$$

It is assumed that there for some α we have $\Theta(\alpha) < \infty$.

Definition 11.1.1 *The zero solution of (1.1) is stable (or, more precisely, B_0-stable) if for any positive ε there is a positive $\delta = \delta(\varepsilon)$ such that the inequality*

$$\|B_0 v_1\|_H + \|v_2\|_H < \delta$$

implies $\sup_{t \geq 0} \|u(t)\|_H < \varepsilon$.

The objective of this chapter is to derive restrictions to the nonlinear term $F(t, u)$ which ensure the stability of the zero solution of (1.1).

11.2   Statement of the result

The main result of the chapter consists in the following

Theorem 11.2.1 *Suppose that conditions (i)-(ii) are valid and the inequality*

$$\theta(\alpha)q < 1 \tag{2.1}$$

holds for some positive α. Then the zero solution of (1.1) is stable. In particular, if the initial data v_1 and v_2 satisfy the condition

$$\|u_1\|_H + \theta(\alpha)[(\|S_0^{-1}W_1 v_1\|_H + |R|\,\|S_0^{-1}W_2 v_1\|_H)$$

$$+\alpha\|S_0^{-1}v_2\|_H] \leq [1 - q\theta(\alpha)]r, \tag{2.2}$$

then the following inequality is true for all $t \geq 0$:

$$\|u(t)\|_H \leq [1 - q\theta(\alpha)]^{-1}\{\|u_1\|_H + \theta(\alpha)(\|S_0^{-1}W_1v_1\|_H$$

$$+|R|\,\|S_0^{-1}W_2v_1\|_H) + \alpha\|S_0^{-1}v_2\|_H\}. \tag{2.3}$$

11.3   Proofs

11.3.1   Main Lemma

Let us consider the non-homogeneous linear equation

$$\ddot{u}(t) + W_1 u(t) - \int_0^t R(t-s)W_2 u(s)ds = f(t), \tag{3.1}$$

where $f(t)$ is a continuous function mapping $[0,\infty)$ into $D(S_0)$.

Lemma 11.3.1 *Suppose that conditions (i), (ii) are fulfilled and there is a positive* α *such that* $\Theta(\alpha) < \infty$. *Then for any* $t > 0$ *a solution* $u(t)$ *of* *(3.1) with the zero initial conditions*

$$u(0) = 0, \quad \dot{u}(0) = 0$$

satisfies the inequality

$$\|u(t)\|_H \leq \theta(\alpha) \sup_{0 \leq s \leq t} \|S_0^{-1}f(s)\|_H. \quad\Box \tag{3.2}$$

Proof: Multiplying (3.1) by S_0^{-1} and applying the Laplace transformation to (3.1), we have

$$\Psi(p)\hat{u}(p) = S_0^{-1}\hat{f}(p), \tag{3.3}$$

where $\hat{u}(p)$, $\hat{f}(p)$ are the Laplace transforms of functions $u(t)$ and $f(t)$, respectively, and

$$\Psi(p) = S_0^{-1}p^2 + \sum_{j=1}^{l_1} A_j^{(1)} \otimes \mu_j^{(1)} - \hat{R}(p)\sum_{j=1}^{l_2} A_j^{(2)} \otimes \mu_j^{(2)}. \tag{3.4}$$

It follows from (3.3) that

$$\hat{u}(p) = \Psi^{-1}(p)S_0^{-1}\hat{f}(p).$$

Employing the properties of the convolution and the inverse Laplace transformation, we can write

$$u(t) = \int_0^t \Phi(t-s)S_0^{-1}f(s)ds, \qquad (3.5)$$

where

$$\Phi(t) = \frac{1}{2\pi i}\int_{-\alpha-i\infty}^{-\alpha+i\infty} \exp(pt)\Psi^{-1}(p)dp$$

$$= \frac{\exp(-\alpha t)}{2\pi}\int_{-\infty}^{\infty} \exp(i\omega t)\Psi^{-1}(-\alpha+i\omega)d\omega.$$

Since $A_{j_1}^{(1)}$, $A_{j_2}^{(2)}$ and S are mutually commuting normal (selfadjoint) operators, they possess common eigenfunctions $\{\psi_m\}$, cf. (Dunford and Schwartz, 1963). For convenience we suppose that $\|\psi_m\|_L = 1$ for any $m = 1, 2, \ldots$.

For any $e \in \mathbf{C}^n$ and any positive integer m the following equality is valid:

$$[S_0^{-1}p^2 + \sum_{j=1}^{l_1}(A_j^{(1)} \otimes \mu_j^{(1)}) - \hat{R}(p)\sum_{j=1}^{l_2}(A_j^{(2)} \otimes \mu_j^{(2)})]^{-1}(\psi_m \otimes e)$$

$$= [\pi_m^{-1}p^2 + \sum_{j=0}^{l_1}a_{j,m}^{(1)}\mu_j^{(1)} - \hat{R}(p)\sum_{j=1}^{l_2}a_{j,m}^{(2)}\mu_j^{(2)}]^{-1}(\psi_m \otimes e). \qquad (3.6)$$

Equations (1.4), (3.4) and (3.6) imply

$$\Psi^{-1}(p)(\psi_m \otimes e) = \Xi_m^{-1}(p)(\psi_m \otimes e).$$

Substitution of this into (3.5) yields

$$\|\Phi(t)(\psi_m \otimes e)\|_H \le$$

$$\frac{\exp(-\alpha t)}{2\pi}\|\int_{-\infty}^{\infty} \exp(i\omega t)\Xi_m^{-1}(-\alpha+i\omega)d\omega\|_{C^n} \|e\|_{C^n} \le$$

$$\frac{\exp(-\alpha t)}{2\pi} \int_{-\infty}^{\infty} \|\Xi_m^{-1}(-\alpha + i\omega)\|_{C^n} d\omega \, \|e\|_{C^n}. \tag{3.7}$$

Recall that due to Corollary 1.2.4 for an arbitrary $n \times n$-matrix Z and an arbitrary regular ζ the estimate

$$\|R_\zeta(Z)\|_{C^n} \le \sum_{k=0}^{n-1} \frac{g^k(Z)}{\sqrt{k!}\rho^{k+1}(Z,\zeta)}, \tag{3.8}$$

where $\rho(Z,\zeta)$ is the distance between the spectrum of matrix Z and point ζ.

The estimate (3.8) implies that

$$\|\Xi_m^{-1}(p)\|_{C^n} \le \sum_{k=0}^{n-1} \frac{g^k(\Xi_m(p))}{\sqrt{k!}\rho_0^{k+1}(\Xi_m(p))}, \tag{3.9}$$

with $\rho_0(Z) = \rho(Z,0)$. It follows from (3.7), (3.9) together with (1.6) that

$$\|\Phi(t)(\psi_m \otimes e)\|_H \le \Theta(\alpha)\exp(-\alpha t)\|e\|_{C^n}. \tag{3.10}$$

Let

$$v = \sum_{m=1}^{\infty} \psi_m \otimes v_m$$

be an arbitrary element of $D(B_0)$. Taking into account that functions ψ_m are orthonormal in Y we obtain

$$\|v\|_H^2 = \|\sum_{m=1}^{\infty} \psi_m \otimes v_m\|_H^2 = \sum_{m=1}^{\infty} \|v_m\|_{C^n}^2. \tag{3.11}$$

It follows from (3.10) and (3.11) that

$$\|\Phi(t)v\|_H^2 = \sum_{m=1}^{\infty} \|\Phi(t)(\psi_m \otimes v_m)\|_H^2 \le$$

$$\Theta^2(\alpha)\exp(-2\alpha t)\sum_{m=1}^{\infty} \|v_m\|_{C^n}^2 = \Theta^2(\alpha)\exp(-2\alpha t)\|v\|_H^2.$$

Therefore, for any $t \ge 0$

$$\|\Phi(t)\|_H \le \Theta(\alpha)\exp(-\alpha t). \tag{3.12}$$

Substitution of estimate (3.12) into (3.5) with the use of (1.7) yields the desired inequality (3.2). □

11.3.2   A linear homogeneous system

Lemma 11.3.2 *Suppose that conditions (i), (ii) are fulfilled and there is a positive α such that $\Theta(\alpha) < \infty$. Then the zero solution of the linear homogeneous equation*

$$\ddot{u}(t) + W_1 u(t) - \int_0^t R(t-s)W_2 u(s)ds = 0 \qquad (3.13)$$

is stable. In particular, the following inequality is true:

$$\|u(t)\|_H \le \|u_1\|_H + \theta(\alpha)[\|S_0^{-1}W_1 v_1\|_H + |R| \, \|S_0^{-1}W_2 v_1\|_H +$$

$$\alpha \exp(-\alpha t)\|S_0^{-1}v_2\|_H]. \qquad (3.14)$$

Proof: First, let us consider the case $v_1 \ne 0$ and $v_2 = 0$. Substitution of the expression

$$w(t) = u(t) - v_1 \qquad (3.15)$$

into (3.13) yields

$$\ddot{w}(t) + W_1 w(t) - \int_0^t R(t-s)W_2 w(s)ds = \tilde{f}(t), \qquad (3.16)$$

where

$$\tilde{f}(t) = -W_1 v_1(t) + W_2 \int_0^t R(s)v_1(s)ds.$$

Applying Lemma 11.3.1 to equation (3.16), we arrive at the formula

$$\|w(t)\|_H \le \theta(\alpha)(\|S_0^{-1}W_1 v_1\|_H + |R| \, \|S_0^{-1}W_2 v_1\|_H).$$

This inequality together with (3.15) implies that

$$\|u(t)\|_H \le \|v_1\|_H + \theta(\alpha)(\|S_0^{-1}W_1 v_1\|_H + |R| \, \|S_0^{-1}W_2 v_1\|_H). \qquad (3.17)$$

We now consider the case $v_1 = 0$ and $v_2 \ne 0$. Applying the Laplace transformation to (3.13) we find

$$\Psi(p)\hat{u}(p) = S_0^{-1}v_2.$$

The Laplace original $u(t)$ has the form

$$u(t) = \int_0^t \Phi(t-s)\delta(s)ds\, S_0^{-1}v_2 = \Phi(t)\,S_0^{-1}v_2,$$

where $\delta(t)$ is the delta function. This equality and (3.12) imply that

$$\|u(t)\|_H \le \Theta(\alpha)\exp(-\alpha t)\|S_0^{-1}v_2\|_H. \qquad (3.18)$$

Since equation (3.13) is linear, inequalities (3.17) and (3.18) imply estimate (3.14). □

Corollary 11.3.3 *Suppose that conditions (i), (ii) are valid and there is a positive α such that $\Theta(\alpha) < \infty$. Then there is a positive constant β such that any solution of (3.13) satisfies the inequality*

$$\|u(t)\|_H \le \beta(\|S_0^{-1}B_0v_1\|_H + \|S_0^{-1}v_2\|_H).$$

Proof of Theorem 11.2.1: Nonlinear equation (1.1) can be considered as a linear non-homogeneous equation (3.15) with $f(t) = F(t, u(t))$. Employing Lemmas 11.3.1 and 11.3.2 we obtain that for any $t \ge 0$

$$\|u(t)\|_H \le \|v_1\|_H + \theta(\alpha)[\sup_{0 \le s \le t} \|S_0^{-1}F(s, u(s))\|_H +$$

$$\|S_0^{-1}W_1v_1\|_H + |R|\,\|S_0^{-1}W_2v_1\|_H + \alpha\exp(-\alpha t)\|S_0^{-1}v_2\|_H]. \qquad (3.19)$$

It follows from (2.2) that $u(0) \in \Omega(r)$. Since function $u(t)$ is continuous, there is an interval $[0, t_*)$ such that $u(t) \in \Omega(r)$ for $t \in [0, t_*)$. For $t \in [0, t_*)$ inequality (3.19) together with (1.2) implies

$$[1 - q\theta(\alpha)]U(t) \le \|v_1\|_H + \theta(\alpha)(\|S_0^{-1}W_1v_1\|_H +$$

$$|R|\,\|S_0^{-1}W_2v_1\|_H + \alpha\|S_0^{-1}v_2\|_H), \qquad (3.20)$$

where $U(t) = \sup_{0 \le s \le t}\|u(s)\|_H$.

According to (2.2), estimation (3.20) can be extended to $[0, \infty)$. Therefore, (2.2) implies (2.3). The B_0-stability of the zero solution follows from (2.3). □

11.4   Example

Following Drozdov and Kolmanovskii (1994, Chapter 4) let us consider two
elastic bars connected by a nonlinearly viscoelastic layer (strip). Suppose
the bars have length l, mass density ρ, cross-section area Ω and moment of
inertia J.

At moment $t = 0$ external loads are applied to the bars. These loads
consist of compressive forces P_1 applied to the ends of the first bar, and
tensile forces P_2 applied to the ends of the other bar.

Under the action of external forces the bars deform. Denote by $u_1(t, x)$
and $u_2(t, x)$ deflections of the bars' at the point with longitudinal coordinate
x at moment $t \geq 0$.

Suppose the bars' deflections are so small that the nonlinear terms in
the expressions for the curvatures can be neglected. Then under a natural
hypothesis, functions $u_i(t, x)$ satisfy the following equations,

$$\rho\Omega\ddot{u}_1 + E_0 J \frac{\partial^4 u_1}{\partial x^4} + P_1 \frac{\partial^2 u_1}{\partial x^2} = -h,$$

$$\rho\Omega\ddot{u}_2 + E_0 J \frac{\partial^4 u_2}{\partial x^4} - P_2 \frac{\partial^2 u_2}{\partial x^2} = h, \qquad (4.1)$$

where E_0 is the Young modulus of the bars and $h = h(t, x)$ is the layer
response. It is assumed that

$$h(t, x) = c\{[u_1(t, x) - u_2(t, x)]-$$

$$\int_0^t R(t - s)[(u_1(s, x) - u_2(s, x)) + \phi(u_1(s, x) - u_2(s, x))]ds\}, \qquad (4.2)$$

where $c = const > 0$ is the foundation rigidity, $R(t)$ is a relaxation kernel,
and $\phi(u)$ is a continuous function satisfying the inequality

$$|\phi(u)| \leq L_\phi |u| \qquad (4.3)$$

with a given coefficient L_ϕ .

Let u_* be the characteristic deflection of the system at the initial moment of time. Introduce the following dimensionless variables and parameters:

$$t_* = \frac{t}{T_*}, \quad x_* = \frac{x}{l}, \quad u_{k*} = \frac{u_k}{u_*},$$

$$\phi_*(v) = \frac{\phi(u_* v)}{u_*}, \quad R_*(t_*) = R(t), \quad T_* = (\frac{\rho \Omega l^4}{E_0 J})^{1/2},$$

$$P_{1*} = \frac{P_1 l^2}{E_0 J}, \quad P_{2*} = \frac{P_2 l^2}{E_0 J}, \quad c_* = \frac{c l^4}{E_0 J}.$$

In the new notation, the relations (4.1) and (4.2) can be written as follows (for simplicity asterisks are omitted):

$$\ddot{u}_1(t,x) + (D^4 + P_1 D^2)u_1(t,x) + c[u_1(t,x) - u_2(t,x)]-$$

$$c \int_0^t R(t-s)[u_1(s,x) - u_2(s,x)]ds = F_1(t,u_1,u_2),$$

$$\ddot{u}_2(t,x) + (D^4 - P_2 D^2)u_2(t,x) - c[u_1(t,x) - u_2(t,x)]+$$

$$c \int_0^t R(t-s)[u_1(s,x) - u_2(s,x)]ds = F_2(t,u_1,u_2), \qquad (4.4)$$

where D denotes differentiation with respect to x and

$$F_1(t,u_1,u_2) = -c \int_0^t R(t-s)\phi(u_1(s,x) - u_2(s,x))ds,$$

$$F_2(t,u_1,u_2) = c \int_0^t R(t-s)\phi(u_1(s,x) - u_2(s,x))ds. \qquad (4.5)$$

For definiteness we confine ourselves to simply supported bars with the boundary conditions

$$u_k(t,0) = u_k(t,1) = 0, \quad D^2 u_k(t,0) = D^2 u_k(t,1) = 0, \quad (k=1,2). \qquad (4.6)$$

Introduce the following operators:

$$B_1^{(1)} = B_1^{(2)} = I, \; B_2^{(1)} = D^2, \; B_3^{(1)} = D^4, \; S = I. \qquad (4.7)$$

It is easy to check that differential operators (4.7) under boundary conditions (4.6) satisfy (i), (ii).

Let

$$u = \begin{bmatrix} u_1 \\ u_2 \end{bmatrix}, \quad F = \begin{bmatrix} F_1 \\ F_2 \end{bmatrix}, \quad \mu_1^{(1)} = \mu_1^{(2)} = \begin{bmatrix} c & -c \\ -c & c \end{bmatrix},$$

$$\mu_2^{(1)} = \begin{bmatrix} P_1 & 0 \\ 0 & -P_2 \end{bmatrix}, \quad \mu_2^{(2)} = \begin{bmatrix} 1 & 0 \\ 0 & 1 \end{bmatrix}. \tag{4.8}$$

By employing expressions (4.7) and (4.8), equations (4.4) can be represented in form (1.1).

It follows from (4.3), (4.5) and the Schwarz inequality that for any $t \geq 0$ we get

$$\|F(t,u)\|_H^2 \equiv \int_0^1 [F_1^2(t, u_1 u_2) + F_2^2(t, u_1, u_2)]dx =$$

$$2c^2 \int_0^1 dx \Big(\int_0^t R(t-s)\phi(u_1(s,x) - u_2(s,x))ds \Big)^2 \leq$$

$$2c^2 L_\phi \int_0^1 dx \Big(\int_0^t R(t-s)|u_1(s,x) - u_2(s,x)|ds \Big)^2 \leq$$

$$4c^2 L_\phi \int_0^1 \Big(\int_0^t R(t-s)z(s,x)ds \Big)^2 dx$$

with the notation

$$z(s,x) = \sqrt{|u_1(s,x)|^2 + |u_2(s,x)|^2}.$$

Take into account that

$$\int_0^1 \Big(\int_0^t R(t-s)z(s,x)ds \Big)^2 dx =$$

$$\int_0^1 \int_0^t R(t-s_1)z(s_1,x)ds_1 \int_0^t R(t-s_2)z(s_2,x)ds_2 dx =$$

$$\int_0^t \int_0^t R(t-s_1)R(t-s_2) \int_0^1 z(s_1,x)z(s_2,x)dx ds_1 ds_2 \leq$$

$$|R|^2 \sup_{0 \leq s \leq t} \int_0^1 z^2(s,x)dx.$$

From this inequality we easily obtain

$$\|F(t,u)\|_H \leq 2cL_\phi |R|.$$

Therefore, inequality (1.2) is fulfilled with $S_0 = I_H$ and

$$q = 2cL_\phi|R|.$$

We have

$$\Xi_m(p) = \begin{bmatrix} X_{11}(p,m) & X_{12}(p,m) \\ X_{21}(p,m) & X_{22}(p,m) \end{bmatrix},$$

where

$$X_{11}(p,m) = p^2 + \pi^4 m^4 - \pi^2 P_1 m^2 + c[1 - \hat{R}(p)],$$

$$X_{12}(p,m) = -c[1 - \hat{R}(p)], \ X_{21}(p,m) = -c[1 - \hat{R}(p)],$$

$$X_{22}(p,m) = p^2 + \pi^4 m^4 + \pi^2 P_2 m^2 + c[1 - \hat{R}(p)].$$

Since $n = 2$, (1.6) can be rewritten as

$$\Theta(\alpha) = \max_{m \geq 1} \frac{1}{2\pi} \int_{-\infty}^{\infty} \frac{1}{\rho_0(\Xi_m(p))}[1 + \frac{g(\Xi_m(p))}{\rho_0(\Xi_m(p))}]_{p=-\alpha+i\omega} \, d\omega.$$

Now we can immediately apply Theorem 11.2.1. The relevant calculations with the relaxation kernel

$$R(t) = \gamma\chi\exp(-\gamma t),$$

can be found in the excellent book by Drozdov and Kolmanovskii (1994, Chapter 4). Here χ is the material viscosity, $0 \leq \chi \leq 1$, and γ is the characteristic rate of relaxation (γ^{-1} is the characteristic time of relaxation). The stability condition in that book is presented in the following form:

$$q < q_{cr} = \sup_{0 \leq \alpha < \gamma} \frac{\alpha}{\Theta(\alpha)}.$$

11.5 Notes

In the last two decades, several publications have been devoted to the study of stability of integrodifferential equations, see e.g. (Burton, 1983), (Gripenberg, Londen, and Staffans, 1990) and the bibliography therein. A significant part of these investigations are stimulated by problems in the viscoelasticity theory cf. (Fabrizio and Morro, 1992), (Renardy, Hrusa, and Nohel, 1987).

The dynamic problem for a linear viscoelastic three-dimensional body has been considered by Dafermos (1970b) using the direct Lyapunov method. This method and a method of linearization were derived by Drozdov and Kolmanovskii (1992) in the stability theory of viscoelastic three-dimensional bodies with finite strains. The method of Fourier transformation was applied to stability problems of a linear three-dimensional body by Fabrizio and Lazzari (1991).

The method of contrasting stability functionals was generalized to the nonlinear problem by Dafermos and Nohel (1979). The problem of asymptotic stability in the viscoelastic one-dimensional theory was discussed by Bloom (1984), Hrusa and Nohel (1985), and MacCamy (1974). In these works some restrictions on nonlinearities were suggested. These restrictions ensure the existence and the asymptotic stability of the zero solution. It was shown that for integral constitutive equations of a viscoelastic material, there exists solution for the dynamic problem in the infinite time interval only for small initial data. This result differs from the similar assertion for differential constitutive equations which states that for a linear viscoelastic material with finite strains there exists a solution over the infinite time interval for arbitrary initial conditions cf. Potier-Ferry (1981).

Other methods for the stability analysis for nonlinear Volterra equations in the finite-dimensional spaces were considered by Clement and Nohel (1981), Hara, Yoneyama, and Itoh (1990), and Hara, Yoneyama and Miyazaki (1992). Some very interesting stability conditions for nonlinear Volterra equations in a Hilbert space were derived by Esquivel-Avila (1992), MacCamy (1977) and Staffans (1979).

Our approach is similar to the method utilized by Fabrizio and Morro (1992). The difference consists in the following. Instead of linear Volterra equations we consider a system of nonlinear Volterra equations, and we employ the Laplace transformation in contrast to the Fourier transformation used in Fabrizio and Morro (1992).

Chapter 12

Semilinear Boundary Value Problems

This chapter is devoted to a class of semilinear operator equations with nonselfadjoint, in general, operators on a tensor product of Hilbert spaces. Conditions for existence and uniqueness of solutions are derived. Estimations for solution norms are obtained. Relevant examples are considered.

In Section 12.1 the notation and preliminary results are collected.

The main result of the chapter is Theorem 12.2.1. It is presented in Section 12.2 and gives solvability conditions for equations on a tensor product of Hilbert spaces. The proofs are presented in Section 12.3.

In Section 12.4 particular cases of Theorem 12.2.1 are considered.

A two points boundary value problem for a second order ordinary differential system on a finite interval is investigated in Section 12.5.

A periodic boundary value problem for a first order ordinary differential system is considered in Section 12.6.

Section 12.7 deals with a boundary value problem for an integrodifferential system.

12.1   Notation

Let H be a tensor product of a Euclidean space $\mathbf{C}^n$ and a separable Hilbert space L, (see Section 6.1).

As above, $D(A)$ and $\sigma(A)$ are the domain and the spectrum of a linear operator A, respectively, $\lambda_k(A)$ $(k = 1, 2, ..,)$ denote eigenvalues of an operator A including their multiplicities, $I, I_L, I_{\mathbf{C}^n}$ denote the identity operators in H, L and $\mathbf{C}^n$, respectively.

Throughout this chapter S is a normal operator in L with a compact inverse one, and the eigenprojects P_k $(k = 1, 2, ...)$(see Section 10.3).

Let $\{W_k\}$ be a sequence of invertible matrices satisfying

$$\max_{k=1,2,...} \|W_k\|_{\mathbf{C}^n} < \infty. \tag{1.1}$$

Define in H an operator $W(S)$ by the formula

$$W(S) = \sum_{k=1}^{\infty} W_k \otimes P_k \tag{1.2}$$

This means that

$$W(S)u = \sum_{j}\sum_{k=1}^{\infty}(W_k h_j) \otimes (P_k y_j).$$

for

$$u = \sum_{j}\sum_{k=1}^{\infty} h_j \otimes y_j$$

with $h_j \in \mathbf{C}^n$ and $y_j \in L$.

The series converges in the norm of H. Due to Lemma 6.3.1 and condition (1.1), $W(S)$ is a bounded operator in H. Put

$$S_0 = S \otimes I_{\mathbf{C}^n}$$

and

$$\Omega(r, S, \nu) \equiv \{h \in D(S_0) : \|S_0^{\nu} h\|_H \leq r\}$$

for a non-negative $\nu \leq 1$ and a positive $r \leq \infty$.

Let us consider in H the equation

$$S_0 W(S)u = F(u), \qquad (1.3)$$

where F maps $\Omega(r, S, \nu)$ into H. We will assume that

$$\|F(h)\|_H \le q \|S_0^{\nu} h\|_H + b \text{ for all } h \in \Omega(r, S, \nu), \qquad (1.4)$$

where q is a non-negative number, and b is a positive one.

A solution of (1.3) is a function $u \in D(S_0)$ satisfying this equation.

Let A be an $n \times n$-matrix. Recall that $g(A)$ and $\gamma_{n,k}$, $k = 0, ..., n - 1$ are defined in Section 1.1. Due to the definition of $g(A)$ we have

$$g(zA) = |z| g(A) \text{ for any complex number } z \qquad (1.5)$$

Corollaries 1.3.7 and 1.3.8 assert

$$g(A) \le \sqrt{1/2} \, N(A - A^*) \qquad (1.6)$$

and

$$g(e^{i\theta} A + z I_{\mathbf{C}^n}) = g(A) \text{ for any real number } \theta \text{ and any complex number } z. \qquad (1.7)$$

12.2   Statement of the results

In order to formulate the main result we denote by d_j the smallest modulus of the eigenvalues of the matrix W_j :

$$d_j = \min_{k=1,...,n} |\lambda_k(W_j)|.$$

Put

$$\mu(W, S, \nu) \equiv \max_{j=1,2,...} \sum_{k=0}^{n-1} |\lambda_j(S)|^{\nu-1} \frac{g^k(W_j)\gamma_{n,k}}{d_j^{k+1}}$$

and suppose that

$$\mu(W, S, \nu)(qr + b) \le r. \qquad (2.1)$$

This condition implies that

$$q\mu(W, S, \nu) < 1. \tag{2.2}$$

Theorem 12.2.1 *Let a sequence $\{W_j\}$ of matrices satisfy (1.1), S be a normal operator in L with a compact inverse one, and F continuously map $\Omega(r, S, \nu)$ into H with the property (1.4) for some $\nu \in [0, 1]$ and $r \leq \infty$. If, in addition, condition (2.1) holds, then (1.3) has at least one solution u. Moreover, the estimation*

$$\|S_0^\nu u\|_H \leq \theta \equiv \frac{b\mu(W, S, \nu)}{1 - q\mu(W, S, \nu)} \tag{2.3}$$

is true.

Remark 1 Theorem 12.2.1 is exact. Indeed, let $F(x) = h$ where h is a known vector. That is, (1.3) is a linear equation in the considered case. Then, we can take $q = 0$ and $r = \infty$. Due to Theorem 12.2.1 for the existence of the solutions it is sufficient that

$$\min_{j=1,2,\ldots} d_j > 0.$$

But this condition is necessary for the existence of solutions of the linear equation.

Remark 2. Let (2.2) hold, then it is simple to check that for any positive b there is a sufficiently large r such that relation (2.1) holds.

Corollary 12.2.2 *Under the hypothesis of Theorem 12.2.1, the estimation*

$$\|S_0 u\|_H \leq (q\theta + b)\mu(W, S, 1)$$

is valid.

Let us suppose that there is a monotonically increasing positive scalar-valued function ψ defined on the positive real semi-axis such that the conditions $\psi(0) \neq 0$ and

$$\|F(h)\|_H \leq \psi(\|S_0^\nu h\|_H) \text{ for all } h \in D(S^\nu) \tag{2.4}$$

are fulfilled. Suppose there exists a (positive) solution of the equation

$$x = \mu(W, S, \nu)\psi(x). \tag{2.5}$$

Theorem 12.2.3 *Let $\{W_j\}$ and S satisfy the hypothesis of Theorem 12.2.1, and F continuously map $D(S_0^\nu)$ into H with the property (2.4) for some $\nu \in [0,1]$. Let the scalar equation (2.5) have a positive solution. Then (1.3) has at least one solution $u \in D(S_0)$. Moreover, it subordinates the inequality*

$$\|S_0^\nu u\| \le x(\psi),$$

where $x(\psi)$ is the smallest (positive) root of (2.5).

12.3 Proofs

Lemma 12.3.1 *Under the assumptions of Theorem 12.2.1, the inequality*

$$\|S_0^{\nu-1}W^{-1}(S)\|_H \le \mu(W, S, \nu)$$

is true.

Proof: Thanks to Theorem 1.2.2, the inequality

$$\|W_j^{-1}\| \le \sum_{k=0}^{n-1} \frac{g^k(W_j)\gamma_{n,k}}{\rho^{k+1}(W_j, 0)}$$

holds. Since $\rho(W_j, 0)$ is the distance from the spectrum of W_j to zero, the equality $\rho(W_j, 0) = d_j$ holds. Therefore, we can write down

$$\|\lambda_j^{\nu-1}(S)W_j^{-1}\|_{C^n} \le \mu(W, S, \nu) \text{ for all } j = 1, 2, \dots .$$

But according to (1.2)

$$S_0^{\nu-1}W^{-1}(S) = \sum_{k=1}^{\infty} \lambda_k^{\nu-1}(S)W_k^{-1} \otimes P_k.$$

Now Lemma 6.3.1 ensures the result. $\square$

Proof of Theorem 12.2.1: Consider the operator Ψ defined on $D(S_0^\nu)$ by the formula

$$\Psi(h) \equiv S_0^{-1}W^{-1}(S)F(h), \; h \in D(S_0^\nu). \tag{3.1}$$

It is well-known (Edwards, 1965) that a product of a compact operator and a continuous bounded mapping is a continuous compact mapping. That is, $S_0^\nu\Psi(h) \equiv S_0^{\nu-1}W^{-1}F(h)$ is a continuous compact mapping.

Exploiting Lemma 12.3.1, we obtain

$$\|S_0^\nu\Psi(h)\|_H = \|S_0^{\nu-1}W^{-1}F(h)\|_H \leq \mu(W,S,\nu)\|F(h)\|_H. \tag{3.2}$$

Now (1.4) gives

$$\|S_0^\nu\Psi(h)\|_H \leq \mu(W,S,\nu)(q\|S_0^\nu h\|_H + b) \tag{3.3}$$

for any $h \in \Omega(r, S, \nu)$.

Due to (2.1) Ψ maps $\Omega(r, S, \nu)$ into itself and therefore it has at least one fixed point due to Schauder's theorem (Edwards, 1965). In other words, the equation

$$u = S_0^{-1}W^{-1}F(u) \tag{3.4}$$

has a solution $u \in \Omega(r, S, \nu)$. Clearly, it belongs to $D(S_0)$. Multiplying (3.4) by S_0, we can assert that (1.3) has a desired solution.

Further, to obtain estimation (2.3) we multiply (3.4) by S_0^ν and take into account (3.3). This yields

$$\|S_0^\nu u\|_H \leq \mu(W,S,\nu)(q\|S_0^\nu u\|_H + b).$$

Hence due to (2.2)

$$\|S_0^\nu u\|_H \leq (1 - q\mu(W,S,\nu))^{-1}b\mu(W,S,\nu).$$

As claimed. $\square$

Proof of Corollary 12.2.2: Thanks to (1.3) and (1.4) the following relation

$$\|W(S)S_0 u\|_H \leq q\|S^\nu u\|_H + b$$

is true, since S_0 commutes with $W(S)$.

It is clear that

$$\|W(S)h\|_H \geq \frac{\|h\|_H}{\|W^{-1}(S)\|_H} \text{ for any } h \in H.$$

Lemma 12.3.1 implies

$$\|W(S)h\|_H \geq \frac{\|h\|_H}{\mu(W,S,1)}.$$

Thus,

$$\frac{\|S_0 u\|_H}{\mu(W,S,1)} \leq \|W(S)S_0 u\|_H \leq q\|S_0^\nu u\|_H + b.$$

Now Theorem 12.2.1 ensures the stated result.

Proof of Theorem 12.2.3: Taking into account that ψ monotonically increases and using (2.5), we can write down $\mu(W,S,\nu) \ \psi(x) \leq x(\psi)$ if $x \leq x(\psi)$. But due to (3.2) and (2.4)

$$\|S_0^\nu \Psi(h)\|_H = \|S_0^{\nu-1} W^{-1} F(h)\|_H \leq \mu(W,S,\nu) \ \psi(\|S_0^\nu h\|_H)$$

Thus,

$$\|S_0^\nu \Psi(h)\|_H \leq x(\psi) \text{ if } \|S_0^\nu h\|_H \leq x(\psi).$$

That is Ψ maps $\Omega(r,S,\nu)$ with $r = x(\psi)$ into itself and therefore Ψ has at least one fixed point due to Schauder's theorem. Consequently, the equation (3.4) has a solution $u \in \Omega(x(\psi),S,\nu)$. Clearly, it belongs to $D(S_0)$.

Multiplying (3.4) by S_0, we can assert that (1.3) has a solution $u \in \Omega(x(\psi),S,\nu)$.□

12.4 Particular cases of Theorem 12.2.1

4.1 Let S satisfy the hypothesis of Theorem 12.2.1. Consider in H the equation

$$S_0 u + cu = F(u), \tag{4.1}$$

where c is a constant $n \times n$-matrix, and $F : D(S_0) \to H$ satisfies the condition

$$\|F(u)\|_H \leq q\|S_0 u\|_H + b$$

$$\text{for } u \in D(S_0) \text{ with } \|S_0 u\|_H \leq r.$$

That is (1.4) is satisfied with $\nu = 1$. Take

$$W_j = I_{\mathbf{C}^n} + \lambda_j^{-1}(S)c. \tag{4.2}$$

Besides,

$$W(S) = \sum_{j=1}^{\infty} (I_{\mathbf{C}^n} + \lambda_j^{-1}(S)c) \otimes P_j = I_H + S^{-1} \otimes c,$$

since

$$S^{-1} P_j = \lambda^{-1}(S) P_j.$$

The result is

$$S_0 W(S) u = (S_0 + c) u.$$

Therefore, (4.1) has the form (1.3). The relations (1.7) and (1.5) imply:

$$g(W_j) = g(\lambda_j^{-1}(S)c) = \frac{g(c)}{|\lambda_j(S)|}.$$

Now (1.6) gives

$$g(W_j) \leq g_0 \text{ for any } j = 1, 2, ..., \tag{4.3}$$

with the notation

$$g_0 \equiv \sqrt{1/2} N(c - c^*) \|S^{-1}\|_L.$$

Besides,

$$d_j = \min_{k=1,...,n} |1 + \lambda_j^{-1}(S)\lambda_k(c)| \geq \overline{d}_0,$$

where

$$\overline{d}_0 = \min_{j=1,2,...} \min_{k=1,...,n} |1 + \lambda_j^{-1}(S)\lambda_k(c)|.$$

Due to (4.3), the relation

$$\mu(W, S, 1) \leq \mu_1(W, S) \equiv \sum_{k=0}^{n-1} g_0^k \frac{\gamma_{n,k}}{\overline{d}_0^{k+1}}. \tag{4.4}$$

is fulfilled. If

$$\mu_1(W, S)(qr + b) \leq r, \tag{4.5}$$

then due to Theorem 12.2.1, equation (4.1) has at least one solution $u \in \Omega(r, S, 1)$, and the inequality

$$\|S_0 u\|_H \leq \theta_1 \tag{4.6}$$

is valid where

$$\theta_1 \equiv \frac{b\mu_1(W, S)}{1 - q\mu_1(W, S)}.$$

4.2. Assume that S satisfies the hypothesis of Theorem 12.2.1, and consider in H the equation

$$(S_0 + \sum_{k=1}^{\infty} T_k \otimes P_k)u = F(u). \tag{4.7}$$

Here P_k are the eigenprojectors of A, again, and $\{T_k\}$ is a uniformly bounded sequence of constant $n \times n$-matrices. The function $F : H \rightarrow H$ satisfies (1.4) with $\nu = 0$. That is,

$$\|F(u)\|_H < q\|h\|_H + b \tag{4.8}$$

for all $u \in H$ with $\|u\|_H \leq r$.

It is clear that (4.7) can be considered as (1.3) with $W(S)$ defined by the formula

$$W(S) = I_H + \sum_{k=1}^{\infty} \lambda_k^{-1}(S)T_k \otimes P_k.$$

I.e. we can take

$$W_k = I_{C^n} + \lambda_k^{-1}(S)T_k \; ; k = 1, 2, \ldots. \tag{4.9}$$

In the considered case

$$d_k = \min_{j=1,\ldots,n} |1 + \lambda_k^{-1}(S)\lambda_j(T_k)|,$$

or

$$d_k = |\lambda_k^{-1}(S)|\rho_k(W, S),$$

where

$$\rho_k(W, S) \equiv \min_{j=1,\ldots,n} |\lambda_k(S) + \lambda_j(T_k)|.$$

Further, according to (1.7) and (1.5)

$$g(W_k) = |\lambda_k^{-1}(S)|g(T_k).$$

Thus,

$$\mu(W, S, 0) = \sup_{k=1,2,\ldots} \sum_{m=0}^{n-1} \frac{\gamma_{n,m} g^m(T_k)}{\rho_k^{m+1}(S, W)}. \tag{4.10}$$

If

$$\mu(W, S, 0)(qr + b) \le r, \tag{4.11}$$

then due to Theorem 12.2.1 equation (4.7) has at least one solution and the inequality (2.3) holds with

$$\theta = \frac{b\mu(W, S, 0)}{1 - q\mu(W, S, 0)}.$$

12.5   A two points boundary problem

In this and in the next two sections we will consider simple examples which only illustrate the applications of Theorem 12.2.1 and by themselves do not claim to be essential.

Let us consider in $\mathbf{C}^n$ the problem

$$d^2 u/dx^2 + cu = F(u, u', x), \quad \text{for } 0 < x < 1 \text{ and } u(0) = u(1) = 0, \quad (5.1)$$

where c is a constant matrix, and the function

$$F : \mathbf{C}^n \times \mathbf{C}^n \times [0, 1] \to \mathbf{C}^n$$

is continuous with respect to all its arguments and satisfies the inequality

$$\|F(h, g, x)\|_{\mathbf{C}^n} \le q_p \|h\|_{\mathbf{C}^n}^p + \overline{q}_{p_1} \|g\|^{p_1} + b \text{ for any } h, g \in \mathbf{C}^n \tag{5.2}$$

with constants $q_p, \overline{q}_{p_1}, b \ge 0$ and $p, p_1 \ge 1$.

The space $H = L^2([0, 1], \mathbf{C}^n)$ is pertinent to our aim, since

$$L^2([0, 1], \mathbf{C}^n) = L^2[0, 1] \otimes \mathbf{C}^n,$$

where $L^2[0,1] = L$ is the space of scalar-valued functions.

A solution of (5.1) is an absolutely continuous function u defined on $[0,1]$ and satisfying the boundary condition and whose second derivative (in the distributed sense) belongs to $L^2([0,1], \mathbf{C}^n)$. That is,

$$u \in D \equiv \{g \in H : g'' \in H; g(0) = g(1) = 0\},$$

and u satisfies (5.1) almost everywhere on $(0,1)$.

It is clear that D is the homogeneous Sobolev space.

Let $G(x,s)$ be the Green function for the scalar boundary value problem

$$d^2 f / dx^2 = 0 \text{ for } 0 < x < 1 \text{ and } f(0) = f(1) = 0.$$

Recall that
$$G(x,s) = \begin{cases} x(s-1) & \text{if } 0 \le x \le s, \\ s(x-1) & \text{if } 0 \le s \le x. \end{cases}$$

Since
$$u(x) = \int_0^1 G(x,s)u''(s)ds,$$

then (by Schwarz's inequality)

$$\|u(x)\|_{C^n} \le \gamma_0 \|u''\|_H$$

and

$$\|u'(x)\|_{C^n} \le \gamma_1 \|u''\|_H \text{ for each } 0 \le x \le 1$$

with the notation

$$\gamma_0 = \sup_{0 \le x \le 1} [\int_0^1 G^2(x,s)ds]^{1/2},$$

and

$$\gamma_1 = \sup_{0 \le x \le 1} [\int_0^1 (\frac{\partial G(x,s)}{\partial x})^2 ds]^{1/2}.$$

These inequalities and (5.2) imply

$$(\int_0^1 \|F(u(x), u'(x), x)\|_{C^n}^2 dx)^{1/2} \le q_p(\gamma_0 \|u''\|_H)^p + \bar{q}_{p_1}(\gamma_1 \|u''\|_H)^{p_1} + b$$

for every $u \in D$. For a fixed positive number r, the latter inequality entails

$$\|F(u(x), u'(x), x)\|_H \le q\|u''\|_H + b \tag{5.3}$$

with

$$q = q_p \gamma_0^p r^{p-1} + \overline{q}_{p_1} \gamma_1^{p_1} r^{p_1-1}$$

if $\|u''\|_H \leq r$.

Set

$$D(S) = \{y \in L \equiv L^2([0,1], \mathbf{C}^1) : y'' \in L, y(0) = y(1) = 0\},$$

and

$$Sy = y'', \text{ for } y \in D(S).$$

That is $D(S_0) = D$.

Take W_j in the form (4.2). Then

$$S_0 W(S)u = u'' + cu = (S \otimes I_{C^n} + I_L \otimes c)u \qquad (5.4)$$

Thus, (5.1) can be written in the form (4.1) and therefore (4.3) holds. Besides, due to (5.3) condition (1.4) is satisfied with $\nu = 1$. Clearly, in the considered case

$$\lambda_k(S) = -\pi^2 k^2 \; ; k = 1, 2, \dots .$$

Thus, $\|S^{-1}\|_L = \pi^{-2}$ and (4.4) holds with

$$\overline{d_0} = \min_{j=1,\dots,n \; ; k=1,2,\dots} |1 - \frac{\lambda_j(c)}{\pi^2 k^2}|.$$

Therefore due to Theorem 12.2.1, the problem (5.1) has under the condition (4.5) at least one solution which satisfies (4.6).

12.6   A periodic boundary problem

Consider in the space

$$H = L^2([0,1], \mathbf{C}^n) \equiv L^2[0,1] \otimes \mathbf{C}^n$$

the problem

$$du/dx + c_0 u = F(u, u', x) \text{ for } 0 < x < 1 \text{ and } u(0) = u(1) \qquad (6.1)$$

with a constant matrix c_0, assuming that the function

$$F : \mathbf{C}^n \times \mathbf{C}^n \times [0,1] \to \mathbf{C}^n$$

is continuous with respect to all its arguments and satisfies the inequality

$$\|F(h,g,x)\|_{C^n} \le q_p\|h\|_{C^n}^p + \overline{q}_1\|g\|_{C^n} + b \text{ for any } h,g \in \mathbf{C}^n \qquad (6.2)$$

with constants $q_p, \overline{q}_1, b \ge 0$ and $p \ge 1$.

A solution of (6.1) is an absolutely continuous function u defined on $[0,1]$ and satisfying the boundary condition with the derivative (in the distributed sense) belonging to $L^2([0,1], \mathbf{C}^n)$. That is

$$u \in D \equiv \{g \in H : g' \in H; g(0) = g(1)\}$$

and u satisfies (6.1) almost everywhere on $(0,1)$. It is clear that D is the homogeneous Sobolev space. Since the periodic differentiation operator is not invertible, put

$$Su = y' + \pi i y, \text{ for } y \in D(S)$$

with

$$D(S) = \{y \in L = L^2([0,1], \mathbf{C}^1) : y' \in L, y(0) = y(1)\}.$$

That is, $D(S_0) = D$.

Take into account that

$$\|u'(x)\|_{C^n} = \|u'(x) - i\pi u(x) + i\pi u(x)\|_{C^n} \le \|S_0 u(x)\|_{C^n} + \pi\|u(x)\|_{C^n}.$$

Then due to (6.2) the inequality

$$\|F(u(x), u'(x), x)\|_{C^n} \le q_p\|u(x)\|_{C^n}^p + \overline{q}_1\|Su(x)\|_{C^n} + \overline{q}_1\pi\|u(x)\|_{C^n} + b. \tag{6.3}$$

is fulfilled. Further, it is clear that the eigenfunctions of S are $exp(2\pi k i x)$ and the eigenvalues of S are

$$\lambda_k(S) = i\pi(2k+1) \; ; k = 0, \pm 1, \pm 2, \ldots . \tag{6.4}$$

Thus, $||S^{-1}||_L = \pi^{-1}$. Any function $h(x)$ from D can be expanded in the Fourier series:

$$h(x) = \sum_{k=-\infty}^{\infty} exp(2\pi kix) \int_0^1 h(s)exp(-2\pi kis)ds.$$

Hence,

$$h(x) = (S_0^{-1}S_0h)(x) = \sum_{k=-\infty}^{\infty} \lambda_k^{-1}(S)exp(-2\pi kix) \int_0^1 (S_0h)(s)exp(2\pi kis)ds.$$

Thanks to the Parseval equality

$$||S_0h||_H^2 = \sum_{k=-\infty}^{\infty} ||\int_0^1 S_0h(s)exp(2\pi iks)ds||_{C^n}^2.$$

Now the Schwarz inequality implies

$$||h(x)||_{C^n} \leq \gamma_3||S_0h||_H \ ; h \in D(S_0),$$

where due to (6.4)

$$\gamma_3 = \sqrt{\sum_{k=-\infty}^{\infty} |\lambda_k^{-2}(S)|} = \pi^{-1}\sqrt{2\sum_{k=0}^{\infty}(2k+1)^{-2}}.$$

Using (6.3), we obtain

$$||F(u(x),u'(x),x)||_{C^n} \leq \overline{q}_1(1+\gamma_3\pi)||S_0u||_H + q_p\gamma_3^p||S_0u||_H^p + b.$$

For a fixed positive number r, the latter inequality entails

$$||F(u(x),u'(x),x)||_H \leq [\overline{q}_1(1+\gamma_3\pi) + q_pr^{p-1}\gamma_3^p]||S_0u||_H + b$$

if $||S_0u(x)||_H \leq r$ .Thus (1.4) is satisfied with $\nu = 1$ and

$$q = \overline{q}_1(1+\gamma_3\pi) + q_pr^{p-1}\gamma_3^p.$$

Put $c = c_0 - I_{C^n}\pi i$ and take $W(s)$ in the form (4.2). Then

$$S_0W(S)u = (S_0 + c)u = u'' + c_0u.$$

Thus, (6.1) takes the form (4.1) and the relation (4.3) is true

Evidently, eigenvalues of W_k are

$$\lambda_j(W_k) = 1 + \frac{\lambda_j(c)}{\lambda_k(S)}$$

or according to (6.4):

$$\lambda_j(W_k) = 1 + \frac{\lambda_j(c)}{i\pi(2k+1)} \quad (k = 0, \pm 1, \pm 2, \dots \; ; \; j = 1, \dots, n).$$

Besides, the relation (4.4) is valid with

$$\overline{d_0} = \inf_{k=0,\pm 1,\dots \; ; j=1,\dots,n} |1 - \frac{i\lambda_j(c)}{\pi(2k+1)}|.$$

Therefore, under conditions (4.5), the problem (6.1) has at least one solution satisfying the estimation (4.6).

12.7 An integrodifferential equation

Consider in the space $H = L^2([0, 2\pi], C^n)$ the equation

$$u''(x) + \int_0^{2\pi} K(x-s)u(s)ds = F(u,x) \tag{7.1}$$

for all $0 < x < 2\pi$; $t \geq 0$ with the periodic boundary conditions

$$u(0) = u(2\pi), \; u'(0) = u'(2\pi), \tag{7.2}$$

where

$$K(x) = \sum_{m=-\infty}^{\infty} exp(imx)L_m$$

and L_m are constant matrices with the property.

$$\lim_{k \to \pm\infty} L_k = 0.$$

Besides, the function F continuously maps $H \times [0, 2\pi]$ into H under the restriction

$$\|F(u,x)\|_H = [\int_0^{2\pi} \|F(u,x)\|_{C^n}^2 dx]^{1/2} \leq q\|u\|_H + b \text{ for all } u \in H \tag{7.3}$$

For example, let

$$F(u,x) = \int_0^{2\pi} Q(u(s),x,s)ds,$$

where $Q(h,x,s)$ is a vector-valued function satisfying

$$\|Q(h,x,s)\|_{C^n} \le q_1(x,s)\|h\|_{C^n} + \frac{b}{\sqrt{2\pi}} \text{ for every } h \in C^n.$$

Here $q_1^2(x,s)$ is a function integrable on $[0,2\pi] \times [0,2\pi]$. Then (by Schwarz's inequality) (7.3) is satisfied with

$$q^2 = \int_0^{2\pi} \int_0^{2\pi} q_1^2(t,s)dtds.$$

Set

$$D(S) = \{y \in L^2[0,2\pi] : y'' \in L^2[0,2\pi] : y(0) = y(2\pi), y'(0) = y'(2\pi)\}$$

and

$$Sy(x) = y''(x) - y(x) \text{ for } y \in D(S).$$

A solution of the problem (7.1), (7.2) is a function $u \in D(S_0)$ satisfying (7.1) almost everywhere. The eigenprojectors of S are

$$P_k h(x) = \frac{1}{2\pi} \int_0^{2\pi} h(s)exp(-iks)ds \; exp(ikx), \; (k = 0, \pm 1, \dots)$$

and the eigenvalues of S are $\lambda_k(S) = -k^2 - 1$; $k = 0,1,2,\dots$. We can write down

$$u''(x) + \int_0^{2\pi} K(x-s)u(s)ds = (S_0u)(x) + u(x) + \int_0^{2\pi} K(x-s)u(s)ds.$$

Evidently,

$$\int_0^{2\pi} K(x-s)u(s)ds = (\sum_{k=-\infty}^{\infty} L_k \otimes P_k)u(x).$$

Since

$$I_H = \sum_{j=0}^{\infty} I_{C^n} \otimes P_j,$$

we have

$$\int_0^{2\pi} K(x - s)u(s)ds + u(x) = \left(\sum_{j=-\infty}^{\infty} T_j \otimes P_j \right) u(x), \qquad (7.4)$$

with $T_j = L_j + I_{C^n}$. Take W_k in the form (4.9) with $\lambda_k(S) = -k^2 - 1$. So

$$W(S) = I_H - \sum_{k=-\infty}^{\infty} T_k \otimes (k^2 + 1)^{-1} P_k$$

or

$$W(S) = I_H - \sum_{j=-\infty}^{\infty} T_j \otimes S^{-1} P_j.$$

According to (7.4)

$$u''(x) + \int_0^{2\pi} K(x - s)u(s)ds = (S_0 W(S)u)(x)$$

and therefore, the problem (7.1), (7.2) can be rewritten as (4.7). Hence (4.10) follows with

$$\rho_k(W, S) = \min_{j=1,\dots,n} |-k^2 - 1 + \lambda_j(T_k)| = \min_{j=1,\dots,n} |-k^2 + \lambda_j(L_k)|,$$

since $\lambda_j(T_k) = \lambda_j(L_k) + 1$ $(k = 0, \pm 1, \pm 2, \dots)$.

Let $q\mu(W, S, 0) < 1$. Then due to Remark 2 (see Section 12.2), there is a sufficiently large r such that condition (4.11) holds. Thus, thanks to Theorem 12.2.1 the problem (7.1), (7.2) under condition (7.3) has at least one solution satisfying (2.3) with $\nu = 0$.

It is clear that we can take S as an elliptic operator and to consider elliptic systems, elliptic integral systems, etc.

12.8   Notes

Basic methods in theory of boundary value problems (the method of upper and lower solutions, the method of Lyapunov-like functions, etc.) are based on the construction of functions which satisfy some inequalities. By these methods many very strong results are obtained (see for instance (Bernfeld and Lakshmikantham, 1974), (Eweritt and Lewise, 1982), (Kiguradze,

1987) and references therein). But finding such functions for systems of differential equations is usually not easy. In particular, the above obtained existence results allow us to avoid the construction of the mentioned functions in appropriate situations.

This chapter is based on (Gil', 1990a, Chapter 5).

Bibliography

[1] Abrahamson, D. L. and Infante, E. F. (1975). A Lyapunov functional for linear Volterra integrodifferential equations, *Quarterly of Appl. Math, 11*: 35-44.

[2] Agathokis, P. and Foda, S. (1989). Stability and the matrix Lyapunov equations for delay systems, *Int. J. Control, 49:* 417-432.

[3] Ahiezer, N. I. and Glazman, I. M. (1981). *Theory of Linear Operators in a Hilbert Space*. Pitman Advanced Publishing Program, Boston, London, Melburn.

[4] Aizicovici, S. (1980). On semilinear Volterra integrodifferential equations, *Israel J. Math., 36 (3/4):* 273-284.

[5] Barmish, B. R. and Shi, Z. (1989). Robust stability of perturbed systems with delay. *Automatica, 25:* 371-381.

[6] Baumgartel, H. (1985). *Analytic Perturbation Theory for Matrices and Operators*. Operator Theory, Advances and Appl., 52. Birkhauser Verlag, Basel, Boston, Stuttgart.

[7] Beauzamy, B.(1988). *Introduction to Operator Theory and Invariant Subspaces,* North-Holland, Amsterdam.

[8] Bernfeld, S. R., Lakshmikantham, V. (1974). *An introduction to Nonlinear Boundary Value Problem* , Ac. Press, New York.

[9] Bhatia, R. (1987). *Perturbation Bounds for Matrix Eigenvalues,* Pitman Res. Notes Math. 162, Longman Scientific and Technical, Essex, U.K.

[10] Bhatia R., Elsner, L. and Krause, G. (1990). Bounds for variation of the roots of a polynomial and the eigenvalues of a matrix. *Linear Algebra Appl. 142:* 195-209

[11] Bloom, F. (1984). On the existence of solutions to nonstrictly hyperbolic problems in nonlinear viscoelasticity. *Applicable Anal. 17*: 115-133.

[12] Boese, F. G. (1991). Delay independent stability of a special sequence of neutral difference-differential equations with one delay, *J. Diff. Eq., 90*: 397-407.

[13] Branges, L. de. (1963). Some Hilbert spaces of analytic functions I, *Proc. Amer. Math. Soc. 106*: 445-468.

[14] Branges, L. de. (1965). Some Hilbert spaces of analytic functions II, *J. Math. Analysis and Appl.*, 11 : 44-72.

[15] Branges, L. de. (1965). Some Hilbert spaces of analytic functions III, *J. Math. Analysis and Appl.*, 12 : 149-186.

[16] Brodskii, M. S. (1971). *Triangular and Jordan Representations of Linear Operators*, Transl. Math. Mongr., v. 32, Amer. Math. Soc. Providence, R1, 1971, MR 41 4283.

[17] Brodskii, M. S. and Brodskii, V. M.(1968). On abstract resentations of linear operators and multiplicative representations to corresponding characteristic functions, *Soviet Math. Doklady , 181* : 511-514.

[18] Brodskii, V. M. (1970). On triangular resentations of operators near to unitary and multiplicative representation their chracteristic functions, *Soviet Math. Doklady , 190*: 510-513.

[19] Brodskii, V.M., Gohberg, I.C. and Krein M.G. (1969). General theorems on triangular representations of linear operators and multiplicative representations of their characteristic functions, *Funk. Anal. i Pril.*, 3: 1-27 (Russian); English Trans., *Func. Anal. Appl.* 3: 255-276.

[20] Burton, T. A., (1983). *Volterra Integral and Differential Equations*, Ac. Press, New York.

[21] Bylov, B. F., Grobman, B. M., Nemyckii V. V. and Vinograd R.E. (1966). *The Theory of Lyapunov Exponents*, Nauka, Moscow.

[22] Casti, J. (1977). *Dynamical Systems and Their Applications*, Ac. Press, New York.

[23] Chiasson, U. (1988). A method for computing the interval of delay values for which a differential delay system is stable. *IEEE Trans. Autom. Control. 33* : 1176-1178.

[24] Christensen, R.M. (1982). *Theory of Viscoelasticity. An Introduction.* Ac. Press, New York.

[25] Clement, P.H., and Nohel, J.A. (1981). Asymptotic behavior of solutions of nonlinear Volterra equations with completely positive kernels. *SIAM J. Math. Anal. 12*: 514-535.

[26] Coppel, W.A. (1978). *Dichotomies in Stability Theory.* Lecture Notes in Mathematics 629, Springer-Verlag, New York.

[27] Corduneanu, C. (1973). *Integral Equations and Stability of Feedback Systems.* Ac. Press, New York.

[28] Corduneanu, C. (1990). *Integral Equations and Applications.* Cambridge Univ. Press, Cambridge.

[29] Courant, R. (1962). *Partial Differential Equations*, Ac. Press, New York , London.

[30] Dafermos, C.M. (1970a). An abstract Volterra equation with applications to linear viscoelasticity. *J. Diff. Eqns. 7*: 554-569.

[31] Dafermos, C.M. (1970b). Asymptotic stability in viscoelasticity. *Arch. Rational Mech. Anal. 37*: 297-308.

[32] Dafermos, C.M. and Nohel, J.A. (1979). Energy methods for nonlinear hyperbolic Volterra integro-differential equations. *Comm. Partial Diff. Eqns. 4*: 219-278.

[33] Daleckii, Yu L. and Krein, M. G. (1971). *Stability of Solutions of Differential Equations in Banach Space*, Amer. Math. Soc., Providence, R. I.

[34] Datko, R., (1986). Not all feedback stabilized hyperbolic systems are robust with respect to small time delays in their feedbacks. *SIAM J. Control Optim. 24*: 152-156

[35] Demidovich, B. P. (1967). *Lectures in Mathematical Theory of Stability* (Russian), Nauka, Moscow.

[36] Diblasio, G., Kunish, K. and Sinestrari, E. (1985). Stability for abstract linear functional differential equations, *Israel. J. Math. 50*: 231-263.

[37] Drozdov, A.D. (1984). Stability of viscoelastic beams on a soft foundation (Russian). *Dokl. Acad. Nauk ArmSSR 79* : 68-72.

[38] Drozdov, A.D. (1985). Stability of rods made of a non-homogeneously ageing material under the conditions of nonlinear creep. *J. Appl. Mech. Techn. Phys. 26*: 592-597.

[39] Drozdov, A.D. and Kolmanovskii, V.B. (1987). Stability of viscoelastic beams on compliant bases. *Mech. Solids 22*: 162-168.

[40] Drozdov, A.D. and Kolmanovskii, V.B. (1992). Stability of equations with persistence describing the dynamics of viscoelastic bodies. *Diff. Eqns. 28*: 161-172.

[41] Drozdov, A.D. and Kolmanovskii, V.B. (1994). *Stability in Viscoelasticity*. Elsevier, Amsterdam.

[42] Drozdov, A.D., Kolmanovskii, V.B. and Velmisov, P.A. (1991). *Stability of Viscoelastic Systems* (Russian). Saratov Univ. Press, Saratov.

[43] Dunford, N. and Schwartz, J.T. (1958). *Linear Operators, part I*, Interscience publishers, Inc., New York.

[44] Dunford, N and Schwartz, J. T. (1963). *Linear Operators, part II* , Interscience publishers, New York, London.

[45] Durikovic, V. (1979a). On a nonlineary stationary parabolic boundary value problem, *Acta fac. rerum natur. Univ. commen. Math., 35*: 55-76.

[46] Durikovic, V. (1979b). On some properties of stationary parabolic mixed problem . *Acta fac. rerum natur. Univ. commen. Math., 35*: 77-94.

[47] Edwards, R. (1965). *Functional Analysis. Theory and Applications*, Holt, Rinehart and Winston, New York.

[48] Elsner, L.(1985). On optimal bound for the spectral variation of two matrices. *Linear Algebra Appl. 71*: 77-80.

[49] Engler, H. (1991). Weak solutions of a class of quasilinear integrodifferential equations describing viscoelastic materials. *Arch. Rational Mech. Anal. 113*: 1-38.

[50] Esquivel-Avila, J.A. (1992). Lyapunov stability properties of second order evolution equations with finite memory. *Nonlinear Analysis. Theory, Meth. Appl, 19*: 701-715.

[51] Eweritt, W. N. and Lewise, R. T. (eds). (1982). *Ordinary Differential Equations and Operators*, Lectures Notes in Mathematics No 1032, Springer-Verlag, New York.

[52] Fabrizio, M. and Lazzari, B. (1991). On the existence and the asymptotic stability of solutions for linearly viscoelastic solids, *Arh. Rational Mech. Anal. 116*: 139-152.

[53] Fabrizio, M. and Morro, A. (1992). *Mathematical Problems in Linear Viscoelasticity*. SIAM Studies in Applied Mathematics, 12, Philadelphia.

[54] Feintuch, A., Saeks, R. (1982). *System Theory. A Hilbert Space Approach*. Ac. Press, New York.

[55] Gantmaher, F. R. (1967). *Theory of Matrices* (Russian). Nauka, Moscow.

[56] Gel'fand, I. M. and Shilov, G. E. (1958). *Some Questions of Theory of Differential Equations* (Russian). Nauka, Moscow.

[57] Gelfond, A. O. (1967). *Calculations of Finite Diferences* (Russian). Nauka, Moscow.

[58] Gil', M. I. (1973a). On operator colligations with nonunitary characteristic operator functions, *Soviet Math. Doklady, 14*: 1035-1037.

[59] Gil', M. I. (1973b). On the representation of the resolvent of a nonselfadjoint operator by the integral with respect to a spectral function, *Soviet Math. Dokl., 14* :1214-1217.

[60] Gil', M. I. (1975). On operator colligations of open systems, *Soviet Math., 19* :19-28.

[61] Gil', M. I. (1979a). Estimating for norms of functions of a Hilbert-Schmidt operator (Russian), *Izvestiya VUZ, Matematika*, 23 : 14-19. English translation in *Soviet Math., 23* :13-19.

[62] Gil', M. I. (1979b). An estimate for norms of resolvent of completely continuous operator, *Mathematical Notes, 26* : 849-851.

[63] Gil', M. I. (1980). On spectral representation for the resolvent of linear operators. *Sibirskij Math. Journal, 21*: 231.

[64] Gil', M. I. (1981a). On global stability of nonlinear nonstationary systems, *Izvestiya of the USSR Ac. Sci. Ser. Technical Cybernetics, 6*: 158-162.

[65] Gil', M. I. (1981b). On stability of nonstationary distributed system, *Automatics and Remote Control 1*: 5-8.

[66] Gil', M. I. (1982). Bounds for the stability domain of quasilinear systems (Russian), *Sibirskij Mat. J. 23*: 219-221.

[67] Gil', M. I. (1983a). On an estimate for resolvents of nonselfadjoint operators which are "near" to selfadjoint and to unitary ones, *Mathematical Notes, 33*: 81-84.

[68] Gil', M. I. (1983b). On an estimate for a stability domain of differential systems (Russian), *Diff. Uravn., 19*: 1452-1454

[69] Gil', M. I. (1983c). On exponent stability domain of diifferential equations, *Differential Eqs. 19*: 1452-1458.

[70] Gil', M. I. (1983d). On a class of absolutely stable systems, *Automatics and Remote Control, 10*: 70-75.

[71] Gil', M. I. (1983e). On one class of systems which are absolutely stable in the Hurwitz angle. *Doklady of the USSR Ac. Sci. 269*: 1324-1327.

[72] Gil', M. I. (1984a). *Operator Functions, Differential Equations and Dynamics of Systems* (Russian), Nauka, Moscow.

[73] Gil', M. I. (1984b). Bounds for the stability domain of control systems, *Izvestiya of the USSR Ac. Sci. Ser. Techn. Cyberenetics, 2*: 187-190.

[74] Gil', M. I. (1984c). On a stability domain of the predator-prey system, *J. of General Biology, 2*: 396-401

[75] Gil', M. I. (1985a). On absolute stability of nonlinear nonstationary distributed systems, *Automatics and Remote Control, 6*: 12-19.

[76] Gil', M. I. (1985b). On a class of absolutely stable control systems, *Doklady of the USSR Ac. Sci, 280*: 811-815.

[77] Gil', M. I. (1985c). Sufficient and necessary conditions for stability of linear systems, *Izvestiya of the USSR Ac Sci. Techn. Cybernetics, 4* : 225-229.

[78] Gil', M. I. (1985d). Two-sides estimates for solutions of nonlinear equations of a class, *Differential Eqs. 21*: 891-894.

[79] Gil', M. I. (1985e). On one class of absolutely stabile distributed systems. *Automatics and Remote Control, 12*: 54-59

[80] Gil', M. I. (1987a). On systems with several nonlinearities which satisfy the Aizerman-Calman hypothesis. *Doklady of the USSR Ac. Sci, 292*: 1315-1318.

[81] Gil', M. I. (1987b). Two-sides bounds for solutions of linear ordinary differential equations, *Differential Eqs. 23*: 2031-2036.

[82] Gil', M. I. (1987c). Conditions for dissipativity and existence of limiting cycles, *Differential Eqs, 23*: 712-714.

[83] Gil', M. I. (1989a). Bounds for solutions of quasilinear parabolic equations, *Differential Eqs. 25*: 723-726.

[84] Gil', M. I. (1989b). Estimates for solutions of nonlinear diffusion systems, *Differential Eqs. 25*: 2082-2090.

[85] Gil', M. I. (1989c) . Stability of essentially nonlinear nonstationary systems, *Soviet Physics Dokl. 34*: 753-755.

[86] Gil', M. I. (1989d) , The freezing method for nonlinear equations. *Differential Eqs. 25*: 912-918.

[87] Gil', M. I. (1990a). *Operator Functions Method in Theory of Differential Equations*, Nauka, Moscow.

[88] Gil', M. I. (1990b) . Operator functions method in the theory of continuous systems, in the book: Borne, P. and Matrosov, V. (eds).(1990). *The Lyapunov functions method and applications*, J.C.Baltzer AG, Scientific Publishing Co. : 69-71.

[89] Gil', M. I. (1991). Perturbation of the spectra of operators of a certain class, *Mathematical Notes, 49*: 328-330.

[90] Gil , M. I. (1992). On estimate for the norm of a function of a quasi-hermitian operator, *Studia Mathematica, 103(1)*: 17-24.

[91] Gil', M. I. (1993a). On inequalities for eigenvalues of matrices, *Linear Algebra and its Applications, 184*: 201-206.

[92] Gil', M. I. (1993b). Estimates for Norm of Matrix-valued functions , *Linear and Multilinear Algebra, 35*: 65-73.

[93] Gil', M. I. (1993c). Estimates for Norm of matrix-valued and operator-value functions, *Acta Applicandae Mathematicae, 32*: 59-88.

[94] Gil', M. I. (1994). A class of absolute stable multivariable systems. *Int. J. Systems Sci., 25:* 613-617.

[95] Gil', M. I. (1995). On absolute stability of Differential-Delay systems, *IEEE, Trans. Automatic Control* (to appear).

[96] Gil', M. I. and Shargorodskay, L. L. (1986a). On one criterion of existence of limit cycles (Russian), *Izvestiya VUZ, Matematika, 30* : 11-14. English translation in *Soviet Math., 30* : 12-14.

[97] Gil', M. I. and Shargorodskay, L. L. (1986b). On oscillations of auto-generators (Russian). *Radiotekhnic and Electronika, 9:* 1818-1822.

[98] Gil', M. I. and Shargorodskay, L. L. (1991). On estimate of norms for resolvents of quasi-Hermitian operator (Russian), *Izvestiya VUZ, Matematika, 35 (1)*: 14-19. English translation in *Soviet Math., 35* : 16-22

[99] Gohberg, I. C. and Krein, M. G. (1969). *Introduction to the Theory of Linear Nonselfadjoint Operators*, Trans. Mathem. Monographs, vol. 18, Amer. Math. Soc., R. I.

[100] Gohberg, I. C. and Krein, M. G. (1970) . *Theory and Applications of Volterra Operators in Hilbert Space*, Trans. Mathem. Monographs, vol. 24, Amer. Math. Soc., R. I.

[101] Gohberg, I. C., Goldberg S. and Kaashoek M. A. (1993) . *Classes of Linear Operators*. Operator Theory. Advances and Applications Vol. 63. Birkhauser Verlag, Basel

[102] Gripenberg, G., Londen, S.-O, and Staffans O. (1990). *Volterra Integral and Functional Equations*, Cambridge University Press, Cambridge.

[103] Halanay, A. (1969). On a nonlinear Volterra equation. *Michigan. Math. J. 16*: 319-324 .

[104] Hale, J. (1977). *Theory of Functional Differential Equations*, Springer-Verlag, New York.

[105] Hale, J. (1989). *Asymptotic Behavior of Dissipative systems*, Mathematics surveys and monographs No 25. AMS Providence, R. I.

[106] Hara, T., Yoneyama, T., and Itoh T. (1990). Asymptotic stability criteria for nonlinear Volterra integro-differential equations. *Funkcial Ekvac., 33*: 39-57.

[107] Hara, T., Yoneyama, T. and Miyazaki, R. (1992). Volterra integro-differential inequality and asymptotic criteria. *Diff. Integral Eqs. 5*: 201-212.

[108] Heard, M.L. (1984). A class of hyperbolic Volterra integrodifferential equations. *Nonlinear Anal. Theory, Meth., Appl. 8*: 79-93.

[109] Henrici, P. (1962). Bounds for iterates, inverses, spectral variation and field values of nonnormal matrices. *Numerishe Meathematik, 4:* 24-39.

[110] Henry, D. (1981). *Geometric Theory of Semilinear Parabolic Equations*, Lecture Notes in Mathematics, No 840.

[111] Hernandez, I. (1986). *Qualitative Methods for Nonlinear Diffusion Equations*, in the book: *Nonlinear Diffusion Problems*, Eds: Fasano, A. and Primicerio, M. Lecture Notes in Mathematics, No 1224, Springer-Verlag, Berlin, Heidelberg Springer-Verlag, Berlin.

[112] Hrusa, W. J. and Nohel, J.A. (1985). The Cauchy problem in one-dimensional nonlinear viscoelastisity. *J.Dif. Eqns, 59*: 388-412.

[113] Hutson, V. (1986). Stability of a reaction-diffusion model of mutualism. *SIAM. J. Math. Anal. 17*: 58-69

[114] Izobov, N. A. (1974). Linear systems systems of ordinary differential equations (Russian). *Itogi Nauki i Tekhniki. Mat. Analis, 12*: 71-146

[115] Karacostas, G. (1987). On a Staffans type inequality for vector Volterra integrodifferential equations, *Colloq. Math. 52*: 313-317.

[116] Kato, T. (1966). *Perturbation Theory for Linear Operators*, Springer-Verlag. New York.

[117] Kiguradze, I. T. (1987). Boundary problems for ordinary differential systems (Russian), *Itogi Nauki i Teknicki*, 30, Nauka, Moscow: 3-103.

[118] Kolmanovskii, V. and Myshkis, A. (1992). *Applied Theory of Functional Differential Equations*, Cluwer Academic Publishers. Amsterdam.

[119] Kolmanovskii V. and Nosov, V. (1986). *Stability of Functional Differential Equations*, Ac. Press, New York.

[120] Konig, H. (1986). *Eigenvalue Distribution of Compact Operators* , Birkhauser Verlag, Basel, Boston, Stutgart.

[121] Krasnoselskii, M. A. (1964), *Positive Solutions of Operator Equations*, P. Noordhoff LTD, Groningen.

[122] Krasnoselskii, M. A. and Pokrovskii, A. (1977). The absent bounded solution principle, *Soviet Math. Doklady, 233*: 293-296

[123] Krasnoselskii, M. A., Pokrovskii A, and Zarubin, A.G. (1983). On a principal bounded regimes absence in absolute stability of distributed parameters systems, *Automation and Remote Control, 3*: 13-19.

[124] Krasnoselskii, M. A. and Zabreiko, P.P. (1984). *Geometrical Methods of Nonlinear Analysis*, Springer-Verlag, New York.

[125] Kress R., de Vries, H. L. de and Wegmann, R. (1974). On nonnormal matrices. *Linear Algebra Appl. 8*: 109-120.

[126] Lakshmikantham, V., Leela, S., and Martynyuk, A.A. (1989). *Stability Analysis of Nonlinear Systems*, Marcel Dekker, Inc, New York.

[127] Lakshmikantham, V., Matrosov, V. M., and Sivasundaram, S. (1991). *Vector Lyapunov Functions and Stability Analysis of Nonlinear Systems*, Kluwer Academic Publishers, Dordrecht, Boston, London.

[128] Levin, J. J. (1963). The asymptotic behavior of the solution of a Volterra equation, *Proc. Amer. Math. Soc. 14*: 534-541

[129] Levin, J. J. (1965). The qualitative behavior of a nonlinear Volterra equation, *Proc. Amer. Math. Soc. 16*: 711-718

[130] Levin, J. J. an Nohel, J. A. (1963). Note on a nonlinear Volterra equation, *Proc. Amer. Math. Soc. 14*: 924-929

[131] Levin, J. J. an Nohel, J. A. (1965). Perturbations of a nonlinear Volterra equation, *Michigan. Math. J. 12*: 431-447

[132] Levitan, B. M. and Sargsjan, I.S. (1991). *Sturm -Liouville and Dirac Operators*, Kluwer Academic Publisher Group, Dordrecht.

[133] Likhtarnikov, A. L. and Yakubovich, V. A. (1983). Absolute Stability of Nonlinear Systems. Appendix to book: V.Razvan *,Absolute Stability of Systems with Delay* (Russian), Nauka, Moscow: 287-355 .

[134] Likhtarnikov, A. L. and Yakubovich, V. A. (1982). Abstract criteria for absolute stability of nonlinear systems with respect to the linear output and their applications I, *Sibirian Math. J., 23 (4)*: 103-121

[135] Likhtarnikov, A. L. and Yakubovich, V. A. (1983). Abstract criteria for absolute stability of nonlinear systems with respect to the linear output and their applications II, *Sibirian Math. J.*, 23 (5): 129-148

[136] Livsic, M. S. (1954). On a spectral decomposition of a linear non-selfadjoint operator, *Math. Sbornik, 34(76)*: 145-199

[137] Livsic, M. S. (1966). *Operators, Oscillations, Waves* (Russian). Nauka, Moscow.

[138] Loginov, A. I., Shulman, B. S. (1988). Invariant subspaces of Banach algebras (Russian). *Itogi Nauki i Tekniki, Mathematical Analysis, 26*: 65-145 , English transl: *Soviet Journal of Mathematics, 26*: 53-130

[139] Londen, S.-O. (1971). The qualitative behavior of solutions of a nonlinear Volterra equation. *Michigan. Math. J. 18*: 321-330.

[140] Londen, S.-O. (1972). On some nonlinear Volterra integrodifferential equations *J. Differential Equations, 11*: 169-179

[141] Lunardy, A. (1985). Asymptotic exponential stability in quasilinear parabolic equations. *Nonlinear Anal. Theory. Meth. and App. 9 (6)*: 563-586.

[142] Macaev, V. I. (1961). On a class of completely continuous operators. *Soviet Math. Doclady. 2*: 972-975.

[143] Macaev, V. I. (1964). On a method of estimating for resolvents of nonselfadjoint operators (Russian) *Dokl. Ac. Nauk SSSR, 154*: 1034-1037.

[144] MacCamy, R.C. (1974). Nonlinear Volterra equations in a Hilbert space. *J. Diff. Eqns. 16*: 373-393.

[145] MacCamy, R.C. (1977). A model for one-dimensional, nonlinear viscoelasticity. *Quart. Appl. Math. 35*: 21-33.

[146] MacCamy, R. C. and Wong, J. S. (1972). Stability theorems for some functional equations, *Trans. Amer. Math. Soc. 164*: 1-37.

[147] Marcus, M. and Minc, H. (1964). *A Survey of Matrix Theory and Matrix Inequalities*, Allyn and Bacon, Boston .

[148] Myshkis, A. D. (1972). *Linear Differential Equations with Delaying Argument* (Russian). Nauka, Moscow, 1972.

[149] Naredra, K. S. and Taylor, J. T. (1973). *Frequency Domain Criteria for Absolute Stability*, Ac. Press, New York and London.

[150] Nohel, J. A. and Shea, D. F. (1976). Frequency domain methods for Volterra equations. *Adv. in Math. 22*: 278-304

[151] Opial, Z. (1967). Linear problems for systems of nonlinear equations, *J. Diff. Equations, 3(4)*: 580-594.

[152] Pao, C. V. (1984). Stabiity of a coupled reaction diffusion system. *Applicable analysis, 17*: 79-86.

[153] Pazy, A. (1981). The Lyapunov method of semigroups of nonlinear contractions in Banach space. *J. Math. Ann. 40*: 239-262.

[154] Philatov, A. N., Sharova, L.V. (1976). *Integral Inequalities and Theory of Nonlinear Oscillations*. Nauka, Moscow.

[155] Phillips, D. (1990). Improving spectral-variation bound with Chebyshev polynomials, *Linear Algebra Appl. 133*: 165-173.

[156] Phillips, D. (1991). Resolvent bounds and spectral variation, *Linear Algebra Appl. 149*: 35-40.

[157] Pietsch, A. (1988). *Eigenvalues and s-numbers*, Cambridge University Press, Cambridge.

[158] Popov, V. M. (1973). *Hyperstability of Control Systems*. Springer-Verlag, Berlin.

[159] Potapov, V. D. (1985). *Stability of Viscoelastic Elements of Structures* (Russian). Strojizdat, Moscow.

[160] Potier-Ferry, M. (1981). The linearization principle for the stability of solutions of quasilinear parabolic equations. *Arch. Rational Mech. Anal. 77*: 301-320.

[161] Potier-Ferry, M. (1982). On the mathematical foundations of elastic stability theory. *Arch. Rational Mech. Anal. 78*: 55-72.

[162] Rasvan, V. (1975). *Stabilitea Absoluta a Sistemelor*. Automate cu Intirziere Ed. Acad. Rep. Soc. Romania, Bucharest.

[163] Reissig, R., Sansone, G. and Conti, R. (1974). *Nonlinear Differential Equations of Higher Order*, Noordhoff International Publishing, Leiden.

[164] Renardy, M., Hrusa, W.J., and Nohel, J.A. (1987). *Mathematical Problems in Viscoelasticity*. Longmans Press, Essex, 1987.

[165] Ringrose, J. R. (1971). *Compact Linear Operators*. Van Nostrand Reinhold Company, London.

[166] Rudin, W. (1964). *Principles of Mathematical Analysis*. McGraw-Hill, New York.

[167] Schmitt, K. (1972). Periodic solutions of nonlinear differential systems. *J. Math. Anal., Appl.*, 40 (1): 174-182.

[168] Shuichi Jimbo and Yoshihisa Morita. (1994). Stability of nonconstant steady-state solutions to a Ginzburg-Landau equations. *Nonlinear Analysis. Theory, Meth. and Appl., 22*: 753-770.

[169] Siljak, D. D. (1978). *Large Scale Dynamic Systems*, North Holland, New York, Amsterdam.

[170] Sleeman, V.D. and Jarvis, R. J. (eds). (1984). *Ordinary and Partial Differential Equations*, Lectures Notes in Mathematics, No 1224, Springer-Verlag, New York.

[171] Staffans, O. G. (1975). Nonlinear Volterra integrodifferential equations with positive definite kernels, *Proc. Amer. Math. Soc. 51*: 103-108

[172] Staffans, O. G. (1977). Boundedness and asymptotical behaviour of solutions of a Volterra integrodifferential equation. *Michigan. Math. J. 24*: 77-95.

[173] Staffans, O. G. (1979). Some energy estimates for a Volterra integrodifferential equation. *J. Differential Equations, 32*: 285-293.

[174] Staffans, O. (1980). On a nonlinear hyperbolic Volterra equation. *SIAM J. Math. Anal. 11*: 793-812.

[175] Stewart, G. W. and Sun Ji-guang, (1990). *Matrix Perturbation Theory*, Academic Press, New York.

[176] Vasil'ev, N. I. and Klokov, Yu. A. (1978). *Theory of Boundary Value Problem for Ordinary Differential Equations* (Russian). Zinatne, Riga.

[177] Vidyasagar, M. (1993). *Nonlinear System Analysis*, second edition. Prentice-Hall.

[178] Vinograd, R. (1983). An improved estimate the method of freezing, *Proc. Amer. Soc. 89 (1)*: 125-129

[179] Weis, D. G. (1974). Note on a nonlinear Volterra integrodifferential system. *Proc. Amer. Math. Soc. 45*: 214-216

[180] Yakubovich, V. A. (1983). Absolute stability of nonlinear distributed parameters systems, *Automation and Remote Control, 6*: 53-61.

List of Main Symbols

351

Index

Milton Keynes UK
Ingram Content Group UK Ltd.
UKHW021636071024
449327UK00020BA/1322